Ecological Criticism for Our Times

Literature, Nature and Critical Inquiry

Ecological Criticism for Our Times

Literature, Nature and Critical Inquiry

Murali Sivaramakrishnan
Ujjwal Jana

AUTHORSPRESS
Publishers of Scholarly Books

Worldwide Circulation through Authorspress Global Network

First Published 2011

by

Authorspress

Editorial

Q-2A Hauz Khas Enclave,
New Delhi-110 016
e-mail: authorspress@yahoo.com

Marketing

E-35/103, Jawahar Park
Laxmi Nagar, Delhi-110 092
e-mail: authorspress@yahoo.com

Ecological Criticism for Our Times: Literature, Nature and Critical Inquiry

ISBN 978-81-7273-588-3

Printed in India at Tarun Offset, Delhi.

Preface

Ecologically sensitive criticism has been practiced and critiqued in isolation by academics and intellectuals in our part of the world and elsewhere for close to two decades now. However, few significant attempts have been made towards any collaborative venture—towards a theory and practice of eco/logical criticism. The present anthology proposes to be the first of its kind. It engages with the logic of ecology and rhetoric of the theory, seeking to inquire into its reach in terms of continuity and interdependence. The history of ecological thinking and the history of literary critical inquiring may have diverse roots, and yet they seek to reinforce the idea of continuity and interdependence—more so in a complex present where all life on planet earth is at stake.

At its inception we had in our minds an anthology of critical pieces focusing on the various aspects of ecological and environmental criticism, its diverse reach and possibilities, globally as well as regionally orientated. But then as things turned out, we had to settle for multiple voices engaging with diverse issues in different locales and issuing from different theoretical positions. Nevertheless the major focus still remains nature and the interface of the human domain. The book could be seen as a sequel to ASLE India's first book on ecocriticism published in 2009 (Nature and Human Nature: Literature, Ecology, Meaning.) The entire process of collection and editing has taken much longer than we expected—apologies are certainly due in this regard. In comparison with the rapidly changing fields of the hard sciences, publication in the humanities as yet remains riddled with crises and thus problematic—funds available are so very little and few, publishers who would venture out in new directions are also few and far between. The sale of printed books the world over has certainly diminished and universities and colleges are the last remaining repositories for these kinds of books! At least this is where critical inquiry still survives. And if at all there is any resistance to the homogenizing forces of globalization it needs to rise only from behind these bookshelves and carrels of unreason!

We invited articles from academics and critics interested and involved in this unique and sensitive critical practice— so very current and equally relevant in our cultural situation. The response was much more than we anticipated. We have selected a group of writings as indicative of the enthusiasm toward and involvement in this exciting and rewarding field of inquiry. They would reveal the rich spectrum of this collective understanding and search for sensitive and meaningful relationship of the human and the non-human world.

Prof. Murali Sivaramakrishnan
Dr. Ujjwal Jana

Contents

1

Introduction: Continuities and Interdependence: Literature, Nature and Critical Inquiry— Ecological Criticism for Our Times

— Murali Sivaramakrishnan

There has been no dearth of critical theoretical thinking in the last century—as a matter of fact the twentieth century witnessed drastic transformations in the mode and manner of theorizing as such, from west, east north and south. The long standing effects of earlier critical enterprises came to be critiqued and inquired into so drastically that the complex geographical landscapes of theories have come to be perceptibly altered and reshaped into a virtually unrecognizable mass—anything is so very possible, and yet so very little is equally actually made possible. Language and meaning have been challenged; the death of the author has been pronounced ever so many times over; the transparency of the mode of literary transaction has been questioned, and the location of the reader within and without the periphery of the textual sphere is now shown to be quite so dubious to the extent of even being virtually non-existent. Race, class, gender, power,position, locale and reading have all come to transect and bisect the visibility and palpability of the literary text so much so that no text appears to make any meaning, for all reading is equally possible and the virtuosity of the text engenders a host of possible misreadings as well. And anyway, once the prison house of language has engaged the human mind there is definitely no beyonding possibility. The endurance of all appears to be the destiny of the critical brain. And yet as one perceptive critic has politely pointed out: language might have trapped us into the cultural constructions of our environment but there is a real world outside which continues to be wounded and harassed critically! So then, speaking *of* nature is also speaking for nature.[1]

Taking its cue from the Eurocentric concentrates, Indian academic theoreticians had dismissed the very idea of nature as culturally constructed that only within a cultural discourse could the living cosmos be framed to make sense. Various feminist and hosts of postcolonial theories have come quite close to promulgating the idea of nature as a critical conceptual tool, but most often shied away from this on account perhaps of the issues of action and its validation that such a view would call for. Writing of nature is one way of not affecting nature but thinking with nature—ontologically and metaphysically. And it certainly goes counter to the contemporary worldview bounded by market economics. To invoke nature appears similar to invoking ancient deities and speaking of the sacred thereby is equated with the retrogressive fundamentalism of a Rousseau-ian anti-progressive renegade in today's world. Nature, ever so silent for so long in human history, does not require the human being to hasten to its aid, but the human being desperately requires to articulate the silent throes of the natural. This is the context of the evolution of ecologically sensitive critical thinking.

Contexts of Ecocriticism

Ecocriticism is a fairly recent entry into the vortex of critical theorizing in Indian academic circles. However ecological concerns and ecological wisdom might not be alien to this antique land where civilization had sprouted so many years ago.[2] In anthropological terms the ideas of culture and nature have been seen as binaries, and as human societies grew more complex and established newer relationships within and without their own boundaries nature came to be relegated to the background. This became more evidenced and pronounced in Western societies probably because of the advent of urbanization as the strategic norm for guiding and forming human development. Culture and civilisation identified the human being in anthropocentric terms as being in constant confrontation with the more primitive forces of nature. Now as Western man advanced into the peripheries of the non-Western world this worldview also travelled with him. This patriarchal, West-centric, development-oriented, materialistic worldview also travelled and commingled with alter/native visions of human advancement and progress in other parts of the globe. This is not to state that materialism and urbanization were essentially Western—modernity and the cultural constructions that it entailed also sprouted in various versions in non-Western contexts, they had their structural connections differently oriented. Nevertheless, European Enlightenment hastened in the renascent ideas of perception of man and nature: the real and the rational came to be mutually interlinked.

As Rene Descartes so eruditely phrased it[3]:

> All science is certain, evident knowledge. We reject all knowledge which is merely probable and judge that only those things should be believed which are perfectly known and about which there can be no doubts.

And as a consequence the contexts of knowledge and its validity came to be more or less rigidly defined: all knowledge needed to be rationally and logically verifiable, repeatable and scientifically evaluated. From this point onwards, perhaps, one can see the inception of excessive belief in the scientific method: source of knowledge became one-pointed and cerebral. The point of its source and organization was the single inner space different from the outer natural. Not that one should take Descartes solely to task for this excessive validation of the rational. However, in more than many ways, his popularized dictum *cogito ergo sum* (I think therefore I am) triggered the direction of the age of Western enlightenment. Nature came to be segregated from the human being and human's being stood alienated and separate from the erstwhile integrated whole. The *cogito* ushered in the dichotomy of the subject and object.

Such scientific observation came to involve the imposition of a European system of classification upon the rest of the world. Pratt[4] summarises the effects of the European-based system of order, what she calls an 'anti-conquest':

> The (lettered, male, European) eye that held the system could familiarize (naturalize) new sites/sights immediately on contact, by incorporating them into the language of the system."

Now more than anything else this essential shift in perspective had effected a drastic consequence, the mechanical view of nature. Further, as the earlier eighteenth century unified sensibility collapsed a new found energizing of the visual sense at the expense of all other senses evolved, and the expansion of Europe's colonies and development of the industrial culture signalled the death of nature as a fount of energy and recuperative spiritual source. The point worth noting here is the domination and spread of a notion of human-oriented anthropocentric (read, male) idea of history and development, the consequences of which have been drastic and detrimental to nature and the human race!

As we have already seen the view of Nature as alienated and apart from the human's being—the Mechanical view of nature—made it amenable to exploitation and manipulation. Knowledge meant the ability to wield power and potentiality, and the ordering of things provided domination and authority. In another dimension of human history the religious systems developed that provided even religious sanction for the domination of nature and the wild/mild.

In many ways the fate of woman had been linked to the fate of nature, and the male values of aggression and dominance were pitted against the feminine values of acceptance and acquiescence. Design and authority went hand in hand. Similarly, as all values other than the rational were secondary and looked upon with doubt, feeling and emotion were considered subjective and irrational and were therefore to be suppressed/repressed. Consequentially all "knowledge" produced through art and literature came to be looked upon with dubiousness in a practical world!

Nature came to be generally equated with wild and wilderness, and culture with the tamed, refined and domesticated. During the Romantic age in Europe there was a violent reaction against this mechanised vision of nature, and the idea of the sublime was resuscitated. However, the dominant idea of nature as wild and something to be tamed and submitted to the culture of "man" has compulsively travelled across the world, and the resistance to this utilitarian view of nature has been looked upon as a deviant aberration—sometimes even as a mere curio. The wretched idea of nature as a mere source and resource for the human being submerged the idea of sacred nature and anyone "wild" enough to sense anything intrinsically valuable in non-human nature was of course "unnatural". Henry David Thoreau wrote: *I went to the woods because I wished to live deliberately, to front only the essential facts of life, and see if I could not learn what it had to teach, and not when I came to die, discover that I had not lived.* (H. D. Thoreau, Walden). And Mahatma Gandhi pointed out: *Earth has enough to satisfy man's need, not enough to satisfy his greed.*

Ecocriticism in the academic form we understand it now perhaps, could trace its inception to the American academies; its watermark being the formation and organization of *ASLE, the Association for the Study of Literature and Environment.* However, I have always held that the idea of ecologically sensitive criticism had evolved over a period of time in the natural intellectual scheme of theorizing, and could be traced to three major springboards. They are (a) the growth and evolution of an ecological conscience (b) the crises in the human sciences (c) a post-theoretical wave. Let us examine these a little closely.

The Rising Environmental Concerns (Ecological Conscience)

Given the wayward march of human civilization in the last few centuries, especially after the industrial age and the age of technology, the emergence of ecological sciences and their current relevance cannot be ignored. In fact, ecology itself had been necessitated primarily because of this drastic consequence of human actions. The depletion of earth's resources and the reduction of natural biodiversities had come to such a pass that human beings had to simply sit up and take stock of their life habits. It is not that human nature interface had not been there in the earlier ages but only that the disastrous consequences became drastic only fairly recently; human development had been in terms of a geometric ratio and the pollution of air, water, earth and the skies thrust themselves into the focus of urgent human attention only fairly recently in earth's history. In all, the rising ecological concern and the emergence of ecological conscience and wisdom could be one of the major reasons why the world of literature itself came to be so infused with this tempo of conservation, preservation, and the recognition of the natural life balance. Literature and art, as is easily perceivable, had always been concerned with the human nature interface; however, the self-reflexive recognition of this nexus is certainly a twentieth century phenomenon, and the academic sanction of this newly emerging worldview came in only through the adoption of ecosensitive critical theories.

So then, the first road to ecocritical theorising has been through this rise of an ecological conscience. Some of the urgent concerns of this factor have been:

Wayward development or *maldevelopment*
Pollution of air, water, earth, and skies
Depletion of earth's resources
Waste management-garbage disposal to nuclear waste
Depletion of biodiversity—unethical dominance of man

Crises in Human Sciences Academia

The excessive growth of theorising in the twentieth century

could be accounted for in many ways; however, the terminal effects of theory and counter theory had been to bracket off physical nature as we have understood it. When linguists and social scientists argued about the trajectory of language and signs, and how we humans have evolved a parallel world of culture and its virtual dimensions, what was at stake was this physical nature. If the world of abstractions that language encompasses were nothing but a collection of socially constructed texts (in the sense of woven fabric) and contexts then the possibilities of encountering the world outside the human mind in its entirety becomes an intellectual impossibility and the human being collapses within his/her own prison house of language! And yet as I have pointed out earlier in this essay, citing a perceptive ecocritic: *language might have trapped us into the cultural constructions of our environment but there is a real world outside which continues to be wounded and harassed critically*! So then among the post enlightenment problematic the trajectory of theorizing in the world has led us directly to engage with physical nature outside the text, and it was a natural extension of the theoretical debate! The text spills over on to the globe!

Post-Theoretical Wave, Post-Feminisms etc.

As argued earlier, the fate of nature had been linked to the fate of woman. Feminist theoreticians have argued endlessly over this issue of interlinking woman and nature. While some like Vandana Shiva state that nature in its essential form as Prakriti is integrally linked to the organic existence of the woman, many other postfeminists argue against this view as nothing but another facile move by the ever-powerful patriarchy to surreptitiously hold back the identity of the woman. We read in Vandana Shiva:

> The shift from *Prakriti* (nature) to natural resources, from Mater to matter was considered (and in many quarters still considered) a progressive shift from superstition to rationality. Yet viewed from the perspective of nature, or women embedded in nature, in the production and

> preservation of sustenance, the shift is regressive and violent. It entails the disruption of nature's processes and cycles, and her interconnectedness. For women, whose productivity in the sustaining of life is based on nature's productivity, the death of *Prakriti* is simultaneously a beginning of their marginalization, devaluation, displacement and ultimate dispensability. The ecological crisis, is at its root, the death of the feminine principle, symbolically as well as in contexts such as rural India, not merely in form and symbol, but also in the everyday processes of survival and sustenance.[5] (*Staying Alive*, p 42)

Either way the rise and development of ecofeminism has served to reveal the significant contributions of feminist thinking to the fostering of a ecologically sensitive criticism inside the academia as well as outside its periphery.

As I have argued elsewhere (see Echoing Ecospiritual Values for a New World, Beyond Borders *The SAARC Journal*, Vol V, No 3 & 4, 2009; 63-70):

> Ecologically sensitive theorizing as it has emerged and developed over the last two decades, has been among the many things an attempt to reintegrate the conceptual and experiential aspect of nature into the mainstream of theory alongside other equally significant aspects of class, race, and gender. As the critical inquiry informed by the mores of cultural studies progressed it naturally became necessary to incorporate the presence and significance of non-human nature at the core of human understanding and to reinstate it at the central position from the peripheral where it had been discarded earlier. Environmental justice movements and eco-critical theory were thus a natural outcome of (among the many other factors leading to and consequent to the several conservation and Green movements) the growing public awareness of profound ecological crisis the world over, as well as the natural culmination of the historical development of contemporary social and critical theory. Of course, the idea of nature has never been one and unchanging, neither monolithic nor homogeneous, and it has thrown open wide uncovered spaces wherein the interplay of nature and human nature could be analyzed, inquired into, critiqued and understood. Truth and

> meaning of life—all very much a part of the grand narratives of a European, post-enlightenment, modernist epoch—would never be segregated from the non human nature and its immanence in human life any more. A trans-human, spiritual value that recognized the intrinsic quality of all existence human and non human (in more than mere Pantheistic sense) would also find its right meaningful place in any ecologically sensitive critical enquiry. (64-65)

And here, more than anything else, the idea of *deep ecology* as developed by Arne Naess is something that reiterates the idea of the sacred at the heart of things. It is an *ecosophy*.

Deep Ecology

According to Arne Naess (1912-2009), "the aim of supporters of the deep ecology movement is not a slight reform of our present society, but a substantial reorientation of our whole civilization." Hence it is an ecosophy. It concentrates on the human relationship with the natural world and supplies a substantial reorientation to a world run astray. A rejection of anthropocentrism. All life on earth has an intrinsic value irrespective of the human angle.

i) *The major tenets of deep ecology:* A rejection of anthropocentrism. All life on earth has an intrinsic value irrespective of the human angle.

ii) Richness and biodiversity are valuable in themselves and humans have no right to reduce this diversity.

iii) An identification with all life.

iv) Caring for the other life forms is part of individual self-realization.

v) A critique of instrumental rationality (emphasis should be not on quantity and efficiency but quality)

vi) Personal development of a total worldview. Individual thinking and action are of utmost significance and later the collective and the social.

This concept of deep ecology is akin to the spiritual. What is aimed at is life enhancing qualitative values very much similar to spiritual enlightenment or artistic fulfilment.

After all, life becomes meaningful only when we start to live fully and selflessly. *Deliberately*, as Thoreau would sanctify the phrase. We need to act in this world recognising the limits to growth and human development—recognising the integrated nature of all and everything: those ethical values that unite the human and the natural environment have to be respected and upheld.

We read in the Upanishads: *iyam prithvi sarvesam bhutanam madhu, asyai prithvyai sarvani bhutani madhu:* this earth is like honey for all creatures and all the creatures are like honey for this earth, *Brhadaranyaka Upanishad* V *brahmanal.* (Cf Mahatma Gandhi's ethical vision—simple living high thinking). We need to recognize and uphold the values and dimensions of self-sufficiency and reject overexploitation of nature and natural resources, keeping in mind the Gandhian dictum that "*Earth has enough to satisfy man's need, not enough to satisfy his greed*!"

New Directions, Possible Expansions

In my earlier book, *Nature and Human Nature: Literature, Ecology, Meaning* (2009), I had pointed out some of the possible new directions that ecocritical theories could aid and abet in our academic curriculum. Let me list them below:

Rediscover our ties with nature

Incorporate scientific temper into English studies—cf the two cultures debate.

Reorient our critical and conceptual tools.

Explore the historical roots of ecocriticism.

Intensive study of our tribal and folk culture—reconnect with our indigenous roots.

Extensive study of environmental movements in other parts of the world.

Recognizing and coming to terms with responsibility.

> Theoretically incorporate nature as a conceptual term
> alongside race, class, gender—
> the global the regional and the local.

It is in this line that the present volume is proffered. That the contemporary critical scene is rife with theories of nature and the environment is a moot point, primarily because such ecosensitive theorizing is the necessary order of the day, in whatever format! The point is that ecologically sensitive criticism has been practised and critiqued in isolation by academics and intellectuals in our part of the world and elsewhere for close to two decades now. However, very few significant attempts have been made towards any collaborative venture, towards a theory and practice of ecological criticism. The present anthology proposes to be the first of its kind. It engages with the logic of ecology and rhetoric of the theory, seeking to inquire into its reach in terms of continuity and interdependence. The history of ecological thinking and the history of literary critical inquiring may have diverse roots, and yet they seek to reinforce the idea of continuity and interdependence—more so in a complex present where all life on planet earth is at stake.

The emergence of *ASLE India,* we believe, is something that serves as a collective platform for such debates. When in the last decades of the last century ASLE was formed in the US, we in India were already practising the very theory but in different and multiple dimensions and in an isolated and sectarian manner. There was no organized methodology there was no collective forum. It was in 2005 that I met Patrick Murphy in Taiwan, and that was at the inaugural conference of ASLE Taiwan. He promptly gifted me with his monumental work: *The Literature of Nature.* I still resort to this massive reference guide during the course of my classes on literature and the environment. In 2004-2006, the European counterpart of ASLE, US was convened at the Alps Adriatic University located in the beautiful water-city of Klagenfurt, Austria. Here I could hold interactions with almost all the world-renowned pioneers and practitioners of ecocriticism. My visit to the

Department of English and Environment at the University of Nevada at Reno as a Fulbright Visiting Scholar that year proffered me a unique opportunity to interact and work with Prof Scott Slovic, Cheryl Glotfelty and Michael Branch among many others. During my lectures and the deliberations I had in the classrooms as well as the host of extramural discussions we had among faculty and research scholars, I benefited a great deal. The international scholars who visited Nevada and California during my period of stay there were all so equally inspiring, creative, as well as erudite and intellectually stimulating. Later I was invited to lecture at the Universities of Salamanca and Extremadura (Caceres) in Spain. During all these international travel and interactions brought me face to face with innumerable ecologically sensitive minds across various cultures. Nevertheless I have been sorely troubled with the fact that there are indeed very few practising scholars in this very significant field. The field is equally fresh and uncontaminated as yet! It was Scott Slovic who proclaimed that among the major aspects of ecocritical practice are these significant issues: *contact*, *communication* and *responsibility*.

I would also like to add these two among them: continuities and interdependence. These essays engage with the problematic of ecocriticism and its cultural and aesthetic dimensions. They deal with a diversity of themes and issues; their major concern is the space of nature—the space that nature affords us and how we engage with that space, physically, psychologically and spiritually.

ASLE India was launched in January, 2005 and I wrote in its opening web page the following as its working manifesto:

> ASLE India is intended as a national forum for the collective understanding of the intimate relations and biological ties existing between literature, art, culture and the environment.
>
> We recognize that environment as a concept, for obvious reasons, cannot be seen separate or segregated from the cultural and natural spheres. And literature by virtue of being what it is—creative, imaginative, emotive and

> expressive—could be seen as engaging with the human mind and the natural world at every point.
>
> The literary and artistic mind of course appears to have perceived this fact in aesthetic terms from prehistoric times, however critical, more specially academic, pursuit of this connection appears to have arisen fairly recently. The now-widespread use of the terms ecological criticism (eco criticism for short), eco-aesthetics, eco-wisdom etc, are to be read as signs in this direction. Of course the rising ecological conscience the world over is one of the main reasons for this.
>
> We recognize the global dimensions of ecological and environmental concerns. Literary and artistic productions by virtue of what they are reflect the deeper cultural preoccupations of a people, and the ecological constitutes a holistic perception. Eco-aesthetics and environmental aesthetics could lead us towards a collective understanding of human and non-human-nature interface.
>
> ASLE India works towards a national awareness of these issues, recognizing fully well the diversities, differences and dissimilarities often inherent in such broader concerns.
>
> ASLE India is intended as a forum for creative interaction between intellectuals, academics, environmental activists, naturalists, nature-lovers, and those involved and committed to these issues, and wish to remain equally open to each other's individual, methodological and disciplinary differences and points of view.
>
> ASLE India hopes also to work in tandem with other similar organizations and collectives.

Our first book *Nature and Human Nature: Literature, Ecology, and Meaning* was published last year and it received wide readership and reviews. This is our second venture in this direction. The range of eccritical theory is spreading and its practitioners are also evergrowing—not only in academic circles but also outside the limited periphery of the intellectuals. Environmental justice movements the world over are raising the conscience of the human nature nexus.

Murali Sivaramakrishnan in the opening essay *Echoing Eco-Spiritual Values for a New World: An Inter-reading of Herman Hesse's Siddhartha* argues that the questions that confronted the thinking minds of the last century—those problems of flesh and the spirit that tormented the inquisitive souls for many generations have simply been safely laid aside as baseless and irrelevant for the youthful spirit of the present day market culture. We, in the present, exist in a world of make-belief that the electronic media churns out and are constantly entertained by the information technology that affords no questions or space for any quests whatsoever. Granted these, how and why are we to re-read a modernist fictional work like Hermann Hesse's *Siddhartha*? What relevance does this text hold for us in the present? Does it hold forth any values for our environmentally sensitive present? The grand questions of life and death, of pain and pleasure, of the body and the soul, the suffering self and the all pervasive nature, all that the book problematises in terms of the Upanishadic questions—Do they vibe with the concerns of the present? Is it once again another Orientalising attempt? Or is it just another attempt to perceive the East as another strange entity and simply cross over? These are among the many questions raised in the text and through an intereading of the text and the discourses that have developed round it over the years, this paper engages the reader and the aesthetic circuit in the process of meaning production with the recent developments in the field of ecological criticism.

Mihai A. Stroe in his monumental essay "Ecomorphic Horizons: From the Dark Ages to the Romantic Big Bang of Culture and Beyond", engages with the entire geography of the changing ideas of the ecological man, *in the history not only of science and literature, but also of religion when perception of man as a being inside the cosmos shifts between different reference points that are themselves fleeting points, fluid frameworks of self-perception.*

To attach the notion of "ecological" to man must thus be in itself a definitory act of science and philosophy at the same time. The evolution of the concept of ecological man in history is otherwise like the history of an iceberg: for us to speak today of an "ecological man" we must already conceptually be standing on the mulscular shoulders of intellectual, philosophical, scientific, religious and artistic giants who laid layer after layer of cognitive energy that in the end ossify and become stable—dynamic structures having in themselves an embedded instability, out of the tension between this stability in instability and instability in stability a pendular vital force of thought and cognition was and is perpetually born, a pendulum of tendencies that fundamentally determine the behaviour of man in the universe. We shall call this basic process the Pendulum of History, as a concept derived from the ideas of Werner Heisenberg and Arthur Koestler, forming a kind of universal cognitive matrix of thought by which man defines himself in relation to everything. Thus, when speaking of "ecological man" we speak of a way of perception, a way of seeing our place in history and the cosmos. Mihai Stroe's territory of inquiry ranges through the Renaissance and Evolutionary Theory, Enlightenment and Deep Ecology, Life Sciences and Atomic Physics. He concludes:

> That we now still live in the shadow of romanticism, as, among others, Harold Bloom and Margaret Atwood pointed out, is only relevant for the present approach by which we mean to attract attention to the reality that the emerging new paradigm of *homo ecologicus* is a necessity if man is to keep on dreaming about a future of his race safe from the spectre of atomic warfare, from "the shadow of nuclear holocaust" which we now are living under, a future that, why not, should draw from the infinite hidden energies of past generations of philosophers, scientists, artists, saints, shamans, etc, all building up an interrelated system of thought with a view to forming a perfect knowledge of life—of the kind today's *ecosophy* is striving to attain, of the kind Ramon Lull had hoped for while thinking to discover the "magnificent art", and Leibniz envisaged while dreaming about the *mathesis universalis*, potentially having the power to revert today's ecocide into a universal feast of life.

B. S. Korde examines the intimate and anxious concerns of Sylvia Plath toward nature and the ecological balance in his *Concerns of Ecology* (Nature) as r eflected in Sylvia Plath's *Crossing the Water* [CW] (1971) and *Winter Trees* [WT] (1971): An Approach:

> Nature as it is understood in this chapter includes ecology in general and a number of phenomena that are parts of ecology. Some of these phenomena include sea, water, ripples on water, rain, rainbows, snow, the sky, clouds, the sun, sunrise, sunset, midnight, the moon, stars, light, dark, horizon, the air, wind, the mountains, hills, rocks, stones, seasons like winter, birds and flies like swans, sparrows, bees, flies, animals like cows, sheep, lamb, snakes, jellyfish, different forms of vegetation like trees, grass, leaves, lilies, blackberries, fruits, reeds, plants, flowers, buds etc. Moreover, some other ecological phenomena like ivory, honeycomb, peasants, gulls, marbles, deserts, orchids etc. are also mentioned in her poems. At times, Plath seems to establish a very close relationship with ecology and its various manifestations so much so that she seems to be treating them almost as her kith and kin. For example, in her poem "Purdah", she writes: "And should/The moon, my/Indefatigable cousin/ Rise:" ...(*WT*, p. 07)
>
> Poets like William Wordsworth, Whitman and Yeats have expressed their desire to go away from the hubbub and crowds of urban settings which disturbs a man's peace of mind (or tranquility). Plath seems to be expressing similar feelings when she says: "Well, one wearies of the Public Gardens: one wants a/ Vacation.../Away from the labeled elms, the tame tea-ros." . (*CW*, p. 50)

Reena J. Andrews in *Ecocentric Dimensions in the Poetry of Robinson Jeffers* examines the deep ecological overtones in his work. In an age of increasing apprehensions about the environment, she argues, poetic voices that question and resist the prevailing views that have contributed to the environmental crisis becomes relevant. The poetry of Robinson Jeffers with his philosophy of inhumanism is an attempt to present an alternative to humanism. His was an effort, at a time when ecology was a newfangled term, to put across an ecocentric vision, to increasingly force our attention towards recognition of our precarious place upon a blighted planet. Throughout his poems is the awareness that mankind's ingenuity has outpaced its wisdom; he attempts to unmask man's universalist claims and points to ecoconsciousness as a corrective to ego-consciousness.

The delusion of progress has resulted in the broken balance between man and his environment. The poet points to the increasing human population and urbanization as factors that upset the equilibrium of nature's web. The present-day ecologists are concerned with these problems and Jeffers's poetic philosophy—which requires identification with the physical world, whereby a deep appreciation of interdependence is acquired-can be related to deep ecological principles and to the Systems theory of life. This chapter examines many of Jeffers's poetic viewpoints that are remarkably congruent with the deep ecological stance: a devotion to place, an insistence on relationships and interdependence, a reverence for the cosmic totality, the legitimising of the non human, the marginalizing of the human element, the concept of an all-pervasive energy, the need to go beyond humanity to a spiritual identification in nature, a view of man as part of a web being destroyed by anthropocentric aggression, a recognition of nature as a process, and a cyclical apprehension of life.

S. M. Gupta in his Ecological Readings of Yann Martel's *Life of Pi*, examines ecocriticism as the study of inter-connectedness between man and environment mediated through gender,class, race, science, culture and economics. *Life of Pi* demonstrates this interrelationship in several ways, he says. The novel is set mainly on the Pacific Ocean. The hero of the novel has been named after a Parisian swimming club Piscine Molitor (Patel); before the shipwreck he lives in the zoo situated in Pondicherry; his father is a zoo keeper; he studies zoology later at the university of Toronto. He holds the view that a zoo is a place of security and plenty for animals. The description of the animal behaviour, flight distances, dislike for invasion on their (animals') territory, zoomorphism and cruelty of man against animals, as presented in the work, as Gupta argues are "insightful". He points out that the whole description can be termed as 'animalus anthropomorphicus'. Martel expounds the notion of nterconnectedness through the natural cycle of 'food chain'. The hyena eats the zebra and the orang utan on the life boat and in turn she is eaten by the tiger. Mr. Pi spends 227 days with a ferocious tiger, and learns how to survive in adverse environment. This shows man's integration with nature. The novel contains vivid and pictorial descriptions of seascape, marine life, weather and carnivorous island. Man spoils the order of this ecosystem by pouring tons of waste in the sea. The novel underscores the interconnectedness between man and nature by suggesting that the mantra for survival is in finding a balance.

Margarita Carretero-González in *The Other Wordsworth*, a female gaze on the natural world closely examines the feminine eye present beside Wordsworth. The simple mention of the surname Wordsworth evokes in the reader a revolutionary conception regarding poetic diction and the role of poetry, together with idyllic images of a Lake District in which immortal daffodils keep forever "fluttering and dancing in the breeze". At William's side there was always another Wordsworth, his sister Dorothy, the poet's invaluable aid and frequent source of poetic inspiration, but also an exceptional woman and talented writer in her own right. The journals that she kept, in her own words, "because I shall give William pleasure by it" have proved a precious tool for the study of William's creative process, since he usually turned to them in order to revive a particular moment before rendering it in verse. Before finding a place in the famous poem, the daffodils had previously "tossed, and reeled, and danced" in Dorothy's journal. Virginia Woolf was one of the earliest voices in praising the vividness in which Dorothy Wordsworth brings to the reader the changing seasons, their effect on her beloved countryside, her brother's swinging moods and the way all this affected her. This chapter explores the way Dorothy

Wordsworth's poetic voice finds its best expression in those parts of her journal in which her brother is displaced from his pervasively central position and his place is taken up by the landscape surrounding Dorothy, which she succeeds in rendering free from the human ego's intervention

Priyadarshi Patnaik, in *Embodying Nature through Aesthetic Experience* attempts to work out an Indian methodology of the environmental aesthetic. Experiencing nature as a work of art has been problematized in the twentieth century and yielded a form of aesthetics which can be broadly called "environmental aesthetics." Traditionally, aesthetic form presupposes an "author," or creator to whose consciousness the aesthetic act can be attributed. In this schema, the movement is from "author" to "art" to "appreciator." It also brings in the notion of "intentionality." Traditionally, an act of art is a conscious and intentional one by its author. For instance, a painter decides what to paint, how and has a reason why s/he performs the act. The work of art and the author are thus linked. The work has a father (mother) or point of origin and in social practice, the author is responsible for the work. Even when a work is anonymous it is a work of human hand and thus still has an author (whose name is unknown). Even when aesthetic stands relegate the author to the background and explore art as an autonomous entity, the function of the artist as author is acknowledge, even if it is thrown to the background. Looking at nature as an aesthetic entity challenges this notion of a creator or author behind the act. Sometimes, it cannot even be called an "act." The act of aesthetic appreciation of nature becomes the moment of both creation and appreciation of the object under scrutiny at the same time. This is a strand which has been explored in the West in recent times—diverse perspectives about how one perceives nature are presented and different theories of aesthetics of nature have emerged. This essay is an attempt to analyze environmental aesthetic with some illustrations and correlations from the tradition of music and painting, in the background of meaning-making strategies that evolve out of meditative tradition and rules of code of conduct of everyday life.

In *People and Forests: An Eco-Social reading of Kadamanitta's poems,* ***P. Rohith***, points out that, unlike the eco-romantics, deep ecologists and spiritual eco-feminists, who try to remedy the ecological crisis through spiritual, biocentric and romantic means, the social ecologist and eco-Marxists try to root the current ecological crisis in the capitalist system. Here the attempt is to uncover the ecological consciousness reflected in the literary compositions with a predominantly environmental concern, produced in Malayalam, the language of Kerala, the south western state— in the decade between 1975 and 1985, written mostly by writer-activists of the movement to protect the *Silent Valley* (the little bit of wet-evergreen rain forest in the south west of the Indian peninsula.) The picture that the Silent Valley forest conjures up is rightly that of a thick, dense, dimly lighted tract of forest with no visible human impact both in the form of settlement and commercial as well as industrial interests. In this aspect, the movement to protect the *Silent Valley* from the proposed hydroelectric project, the paper argues, differs immensely from that of other environmental struggles like Narmada, Chipko, Mavoor and Baliyapal which evidently rely on the rhetoric of social justice and humanitarianism. Interestingly enough, the conspicuous absence of the human element was not a deterring factor that distracted literary interest. Writers' interest in the Silent Valley was not restricted to the ecological significance of the area, rather they were drawn to it for its latent eco-social critiquing possibilities. This position might appear paradoxical for the protection of the Silent Valley employed, directly or otherwise, no vocabulary of philanthropy and local/regional justice. Rohith attempts to analyse what he terms, *eco-marxist poems* written around the theme of the Silent Valley, with special emphasis on Kadammanitta Ramakrishnan's poetic contributions to analyse the evocative spirit, the revolutionary preoccupations, futurity, pro-Naxal political stance and diction in its attempt to critically locate this streak in the canon of eco-writing in Kerala, which leads on to a study in comparison with/ in contrast on the rhetoric of such movements elsewhere.

Elmar Schenkel in his *Exploring the Physical Roots of Architecture: Poppo Pingel and Hugo Kükelhaus,* points out that

Architecture requires *philosophical inspiration*, otherwise it becomes flat and stale and, as he says, those who inhabit will have to suffer. One of Poppo's inspirations has been the German thinker Hugo Kükelhaus, whose work is hard to define. Schenkel asks: *was he an architect, a philosopher, a biologist, a carpenter, a designer, a painter or an ecologist*? He answers: *He was all of that and more, one of the last flickers of the uomo universale, the polymath of Renaissance days, an afterthought of Goetheanism in the 20th century*. Goethe certainly was his model in so far as he connected art and science, observation and practice. He designed schools and swimming pools for children and handicapped people, collaborated with architects and developed toys for babies. He designed church windows and furniture, painted walls and created herbal gardens. He was on a permanent lecture circuit on which he inspired thousands and encouraged them to use their own sensory experience. He wrote books on inhuman architecture and on the role of the body for buildings. He also developed what was to become fields of sensory experience (*Erfahrungsfelder für die Sinne*) which are now to be found in places all over Germany and Switzerland. Kükelhaus liked to quote Goethe in all kinds of situation. His written works abound with quotations and his whole outlook points to the great classical writer and thinker. In his sensory field, he put up a station at which one can explore Goethe's colour theory. When, on one occasion, journalists asked him how he had "hit upon Goethe", he replied: "You had better ask how Goethe hit upon me!" This was considered an arrogant answer, which in fact it is not. It is the opposite. Goethe, if we follow this line of thought, had covered everything, from poetry to the intermaxillary bone, and thus had also covered his followers, had foreseen Kükelhaus. Be that as it may, the Goethe quotation Kükelhaus gave to Poppo ended up in India. It was translated into Gujarati and now decorates the *chaikhana* of Poppo's and Mona's house in *Auroville*, near Pondicherry in Tamil Nadu, South India.

Usha V.T in her essay *Many Ecologies of the Self: Inner and Outer Spaces in the Poetry of Kamala Das / Suraiya, An Ecofeminist Reading,* maintains that contemporary critical readings of Kamala Suraiya's work albeit the fact that they have emphasized various aspects of her poetry and poetics through an equally large number of learned dissertations and critical expositions, however, have failed to highlight the ecocritical aspects of her poetry and poetics. This attempt to read her poetry from an ecofeminist point of view, Usha claims, traverses new grounds and helps provide new insights into hitherto unexplored regions in her poetic world such as the connections between the human and the non-human world of, her celebration of the body, and male-female dichotomies. This essay attempts to draw attention to the blatant violation of conventional hierarchies through Kamala Das/Suraiya's poetic voices and the placing of her poetic personae within their specific ecological environment which make her poetic output *ecofeminist*. In addition it attempts to identify her capacity to integrate binary oppositions and speak from within alternate viewpoints. The prime persona in her poems is *herself*, which she locates either in a physical environment or in relation to other personae.

K. Reshmy in her *Body as Colony: An Ecofeminist Analysis of Margaret Atwood's The Handmaid's Tale* argues that exploitation and victimization of woman and nature has been a dominant theme in Margaret Atwood's writings. Her famous novel *Surfacing* could be considered an ecofeminist treatise as such. However, it is in *The Handmaid's Tale*, that Atwood graphically explains the horrible consequences of patriarchal and capitalistic colonization over women and nature. Science, with its reductionist policy, has indiscriminately exploited nature for man's benefit. Deforestation, industrialization, hazardous nuclear testing and environmental pollution have far reaching consequences on the health of the people and the biosphere. In *The Handmaid's Tale*, Atwood points out how such a situation can affect women more than men. Atwood discusses how female body can become a territory to be colonised by the patriarchy, how female sexuality and reproduction can come to be regulated by men. In such a system, woman, like nature, becomes a national resource. Atwood employs ecofeminist metaphors to emphasise the theme of oppression of woman and nature and makes the colonised female body the body of the narrative.

S M Gupta in his *Ecological Readings of Yann Martel's Life of Pi*, examines ecocriticism as the study of interconnectedness between man and environment mediated through gender,class, race, science, culture and economics. "Life of Pi" demonstrates this interrelationship in several ways, he says.The novel is set mainly on the Pacific Ocean. The hero of the novel has been named after a Parisian swimming club Piscine Molitor (Patel); before the shipwreck he lives in the zoo situated in Pondicherry; his father is a zoo keeper; he studies zoology later at the university of Toronto. He holds the view that a zoo is a place of security and plenty for animals. The description of the animal behaviour, flight distances, dislike for invasion on their (animals') territory, zoomorphism and cruelty of man against animals, as presented in the work, as Gupta argues are "insightful". He points out that the whole description can be termed as 'animalus anthropomorphicus'. Martel expounds the notion of nterconnectedness through the natural cycle of 'food chain'. The hyena eats the zebra and the orang utan on the life boat and in turn she is eaten by the tiger. Mr. Pi spends 227 days with a ferocious tiger, and learns how to survive in adverse environment. This shows man's integration with nature. The novel contains vivid and pictorial descriptions of seascape, marine life, weather and carnivorous island. Man spoils the order of this eco-system by pouring tons of waste in the sea. The novel underscores the interconnectedness between man and nature by suggesting that the mantra for survival is in finding a balance.

Sumana Biswas in *An Ecocritical Expedition into Shakespeare's Cymbeline* attempts not only to assess the relationship between human beings and Nature but also to ensure, whether civilization's roles conform to the principles of ecology in order to maintain a harmonic and balanced eco-human relationship. This study goes on to highlight the intertwined identity, role and position of women and Nature. It also pictures how the play has critiqued civilization's indiscriminate acquisitiveness of the material world with an egotistic approach that makes it incognizant of the real power and quintessence of the enigmatic Nature. Its is Sumana's view that the ecocritical pursuit equally projects the consumerist civilization's propensity to overexploit and commodify Nature and the 'weak others' with paucity of interest and initiatives for their conservation and rehabilitation.

An ecocritical survey of *Cymbeline* also projects a mosaic of certain other issues as reinforced in the essay: the practical and symbolic significance of Nature in human lives, the impact of Nature upon civilization and vice-versa, the benevolence and vengeance of Nature, anthropocentrism practiced against Nature's law, which in turn nullifies and dismantles it repeatedly; elevated and distorted symbolic association of humans with the non-humans, the rhythms of dynamic Nature in the context of ever changing society and the inconsistent human mind, and the metamorphosis of ecocidal entities into eco-friendly agents brought about by the forces of Nature.

Wanda Baxter's essay considers the introduction of an undergraduate course that views 'the environment' from different perspectives of observation and interpretation. The purpose of the course is to introduce students to interdisciplinary expressions of environmental observation, in order to gain appreciation of the interdependence of creative and scientific ways of knowing, especially as they pertain to environmental sustainability. Wanda Baxter holds that the separation of literature and the humanities from the sciences - in theory and in practice – is part of what limits our ability to respond to environmental problems. Without being able to observe and/or tell the story of a place and its people, science attempts to respond to environmental degradation via linear, detail-oriented means. What we need now, she states, is education that teaches the *whole of a place*. Our world is complex, and our teaching and learning curriculum needs to step up to the challenge of teaching complex ideas. This essay introduces the idea of 'applied literature to environmental issues'. The exercise is ecocritical at the core and its intent is practical. The role of the imaginative thinker and writer needs to be reclaimed as an essential part of understanding and responding to our world.

Ujjwal Jana's *Representation of Nature in Ancient Indian Literature: Some Reflections*, focuses on his basic premise that in Indian world view the awareness of the harmonious bond of the human nature with the physical environment was already acknowledged and a survey of the classical literature speaks not merely of the poets' profound preoccupation with, but also the touching sensibility to the entire gamut of biosphere. He argues that In Indian philosophical thought, the organic relation of humans to the biological nature has always been emphasized. That nature is an inseparable part of human and cultural discourse has been articulated in the *Vedas*. "...everything that exists is an emanation from and is, in essence, Brahman (absolute) and this leads Indian philosophers to use the language of ' Man is Nature', 'Man in Nature' and not 'Man against Nature'.

Life Lines :Water, Life, and the Indian Experience—Cultural Meanings, Social Significance and Literary Implications, attempts to examine the socio-cultural significance and the imaginary implications of water—as *vital element, aesthetic metaphor, image* as well as *symbol*. As ***Murali Sivramakrishnan*** argues, very much like its fluidity and reflective transparency, this vital element, no doubt, plays a significant role in human imagination, and socially constructed reality alike. It conceals and reveals at the same time. Bereft of water there is no life, nevertheless, too much of it hastens human casualty. Drought and flood, famine, starvation and tsunami—all constitute the condition of human life on earth. This essay is divided into two parts: the first is of a general kind that seeks to examine the social and spiritual significance of water in the multi-planed Indian context, while the second is of a textual kind focusing on a work of literature—Ashokamitran's *Water*— that aims to *problematise* cultural and historical meanings and the literary implications of water.

It is not unusual for editors these days to compile various views expressed by a variety of authors with little or no involvement of their own in the bargain! However, contrary to these practices I have contributed three essays to this anthology: this introductory essay is expected to position the volume in the contexts of eco-critical theory and practice, while the one on Herman Hesse is an attempt at an exploration of certain mislaid spiritual values which have been swept under the carpet in the wake of post-structuralism and its discourses. The third one makes an effort in the direction of *problematising* cultural and historical meanings.

This volume is, on the whole, the result of a combined effort of diverse hands. And it echoes multiple views. There is definitely a common thread—that of nature and human nature, however, there are bound to be differences in perspectives. Each essay is necessarily the view and perception of its author. Well, after all, it is through differences and diversification that we may seek our own selves—it is never through a process of homogenization and unification!

Notes and References

1. "In short, it is not language which has a hole in its ozone layer; and the real thing continues to be polluted and degraded even as we refine our deconstructive insights at the level of the signifier." [Kate Soper, What is Nature? Oxford: Blackwell, 1995; 151]
2. It would be interesting to note that the earliest known records of measures taken for the preservation of non-human life forms comes from this part of the world: Emperor Asoka's Fifth Pillar Edict by which game and fishery laws were introduced in the 3rd Century BCE.
3. Descartes quoted in. Hooker, Michael, ed. *Descartes: Critical and Interpretive Essays*. Baltimore: Johns Hopkins Press, 1978
4. Pratt, M. Imperial Eye. London: Routledge, 1992:31 quoted in "Sensing Nature" in Macnaghten, Phil and John Urry, *Contested Natures*. London: SAGE, 1998:112
5. Vandana Shiva, *Staying Alive*, p 42.

Works Cited

Hooker, Michael, ed. *Descartes: Critical and Interpretive Essays*. Baltimore: Johns Hopkins Press, 1978.

Murali , Sivaramakrishnan Ed. Nature and Human Nature: Literature, Ecology, Meaning New Delhi: Prestige, 2009.

Murphy, Patrick. Ed. *The Literature of Nature: An International Sourcebook.* Chicago and London: Fitzroy Dearborn Publishers, 1998.

Pratt, M. Imperial Eye. London: Routledge, 1992:31 quoted in "Sensing Nature" in Macnaghten, Phil and Urry, John. *Contested Natures*. London: SAGE, 1998:112

Shiva, Vandana. *Staying Alive*: *Women, Ecology and Development in India*. New Delhi: Kali for Women, 1988.

Slovic, Scott. *Seeking Awareness in American Nature Writing: Henry David Thoreau, Annie Dillard, Edward Abbey, Wendell Berry, Barry Lopez.* Salt Lake City: Univ. of Utah Press, 1992.

Soper, Kate. What is Nature? Oxford: Blackwell, 1995.

Thoreau, H D. *Walden*, (1854). Boston: Beacon, 1997.

2

Echoing Eco-Spiritual Values for a New World: An Inter-reading of Herman Hesse's *Siddhartha**

— Murali Sivaramakrishnan

Ah, how glorious!
Green leaves, young leaves
Glittering in the sunlight.
—Basho *on visiting the shrine on Mount Nikko*

The moving finger writes and having written moves on...
—Omar Khayyam, *Rubaiyat*

Above all, he learned ...how to listen, to listen
with a heart, with a waiting, open soul,
without passion, without desire, without
judgement, without opinions...
—*Hesse,* Siddhartha, *p.85*

We in the present live in a world devoid of any spiritual qualities, a world of objects and things drained of any value even remotely akin to the sacred. Our contemporary technological modernity or its portentous variations of many "posts" that has hastened us unwittingly into a world of commodities and prices has perhaps blinded our eye to any such levels beyond the commonplace. As one extended *globalised village* we all are made to think proudly about our present age of information and communication. Life styles are becoming more or less similar in almost all parts of the world; there is hardly any difference! What else do we need? At the end of the twentieth century after the disappearance of the Soviet bloc that signalled the final triumph of capitalist ideology there were also intellectual voices speaking from the developed world of *the end of history* and what not! And now we have entered the era of globalisation that magically enchants us into submitting wilfully to its conditioning hegemony as mere unthinking humanoids. There is now no more need for the great *grand narratives* of the past—they have all been revealed to us as being mere intellectual and ideological pretensions of a lost homogenizing and totalising culture. It is as if the contemporary technocratic culture, of course, obviously lays claim to nothing of this sort! The questions that confronted the thinking minds of the last century—those problems of flesh and the spirit that tormented the inquisitive souls for many generations—have been safely laid aside as baseless and irrelevant for the youthful spirit of the present day market culture. We currently exist in a world of make-belief that the electronic media churns out and are constantly entertained by the information technology that affords no questions or space for any quests whatsoever. Granted these, how and why are we to re-read a modernist fictional work like Hermann Hesse's *Siddhartha*? What relevance does this text hold for us in the present? Does it hold forth any values for our environmentally sensitive present? The grand questions of life and death, of pain and pleasure, of the body and the soul, the suffering self and the all pervasive nature, all that the book problematises in terms of the Upanishadic questions—Do they vibe with the concerns of the present? Is it once again another *Orientalising* attempt? Or is it just another attempt to perceive the East as another strange entity and simply cross over?

The march of history after the European Enlightenment has been virtually one sided in more than one sense. The river of inquiry of human significance and the quest for meaning had come to be visualized with but one river *bank*: there was no significant *other*! The Western paradigm of progress, and human development through the hegemonial discourse of science and its handmaiden technology came to symbolize the ordered rationalised vision for all mankind as a whole! The discourse of course produced its own other within its Western paradigm that worked in two ways: while the West dominated and colonized the East it created and narrated the idea of the *Orient* (Edward Said and *Orientalism*) and denigrated the entire non-Western discourses as secondary, derivative and subordinate to the rational real of the West and its hedonistic modernity. Consequently it also created a desire in the minds of many western scholars discontented with the march of Western history to seek in the Romantic East the many answers to their wounded civilization. The nineteenth century search for alternate grand narratives to subvert the over-dominant arrogance of Eurocentric beliefs resulted also in some of the most sensitive and perspicacious readings of Eastern religious and spiritual texts and ways of living and thinking. Academic and non-academic space for clear-sighted, transparent, and intelligible debates were created and the sensitive reader could glean some home-truths of the *spirit of the East*. Hermann Hesse belongs to this struggle to reach the third bank of human longing. In his never-ending search for the alternate model for history's waywardness he turned to the mystic East. However, his profound yearning for the third bank of the river of being was never one of *othering* or domination, but one of silence, understanding, and reverence.

Siddhartha is a unique piece of evidence of a genuine search for peace and human compassion. It is a narrative neither conditioned by the logic of modernity's excess nor bound by the discourse of Western dominance. It is a classic of compassion that groups itself beside non-canonical narratives that proffer unmediated narration and unprejudiced clarity of vision. It goes beyond the limiting rationality of the thinking West in a genuine search for eternal truths. What are these eternal truths? Are we here debating again the monolithic stature of a single unique vision that does not permit the space for even the second bank of the river? Or are we in search of a *unifying vision*, not an already made-up one, but something that neither segregates or homogenizes—a holistic plurality that easily accommodates oneness and differences, that reaches beyond all conditions and historicity? Perhaps there is the third bank of the river, after all? Towards the end of the book, when Siddhartha sits besides the river attentively, we read:

> *Siddhartha listened. He was now listening intently, completely absorbed, quite empty, taking in everything. He felt that he had now completely learned the art of listening. He had often heard all these before, all these numerous voices in the river, but today they sounded different.... And all the voices, all the goals, all the yearnings, all the sorrows, all the pleasures, all the good and evil, all of them together was the world. All of them together was the stream of events, the music of life.*(107)[1]

Hesse's Siddhartha is a unique individual, a sensitive soul who embodies the questing mind of the West and the compassionate soul of the East. Suffering is a common factor, but its reception and the human response radically differ. The Buddha did not attempt to interpret or explain the world, because his goal was salvation from suffering. For Siddhartha the answer lay elsewhere.

Ecologically sensitive theorizing as it has emerged and developed over the last two decades, has been among the many things an attempt to reintegrate the conceptual and experiential aspect of nature into the mainstream of theory alongside other equally significant aspects of class, race and gender. As the critical inquiry informed by the mores of cultural studies progressed it naturally became necessary to incorporate the presence and significance of non-human nature at the core of human understanding and to reinstate it at the central position from the peripheral where it had been discarded earlier. Environmental justice movements and eco-critical theory were thus a natural outcome of (among the many other factors leading to and consequent to the several conservation and *Green* movements) the growing public awareness of profound ecological crisis the world over, as well as the natural culmination of the historical development of contemporary social and critical theory. Of course, the idea of nature has never been one and unchanging, neither monolithic nor homogeneous, and it has thrown open wide uncovered spaces wherein the interplay of nature and human nature could be analyzed, inquired into, critiqued and understood. Truth and meaning of life—all very much a part of the grand narratives of a European, post-enlightenment, modernist epoch—would never be segregated from the non human nature and its immanence in human life any more. A trans-human, spiritual value that recognized the intrinsic quality of all existence human and non human (in more than mere Pantheistic sense) would also find its right meaningful place in any ecologically sensitive critical enquiry. Works like *Siddhartha*, albeit engendered in a modernist discourse, could certainly yield new meanings when re-read under the rubric of ecological critical theory. Indeed, *Siddhartha*, does more than that—it calls for a reordering of the value systems of the market-oriented contemporary present. This is where eco-spirituality and an eco-dharma would invoke us to pause for a meaningful moment in our delirious quixotic reaching after the strange gods of unreason. This is not fictionalization of life, but on the other hand, a reflection of the truth as it should be or ought to be, which in turn would make our contemporary life into the fabric of the fictional. Siddhartha's search for the meaning of his life extends into the present and he moves among us as more real than the virtual images of our selves that the consumer culture of the present has reduced us to.

There is little doubt that *Siddhartha* remains among some of the greatest works produced in the twentieth century—for its simplicity and poignancy of narrative, for its resolute and unwavering pursuit of the position of the thinking individual in a changing world, for its profundity of vision and insights, for its deep commitment to clarifying and questioning the spiritual and the corporeal at the heart of human existence, and above all for its being what it is: *a genuine human tale of self and soul and their intriguing dialogues*. Now, having said this we need to ask ourselves how the novel manages to blend successfully so much philosophical inquiry into its tiny canvas of a little over one hundred pages. The form itself is like narrating a fable that unfolds spontaneously in to the three levels of body, mind and soul, and self-realisation in every stage of being. Hesse has set his tale somewhere in the *mystical East* (this adds to its strangeness of setting and provides the reader with a whole arena of different discourses of the flesh and spirit, of passion and sympathy) without making it out as an *Orientalist narrative* and this serves to provide the story with the sufficient distance from life and suitable involvement with it at several levels at the same time.

Siddhartha is the Brahmin's son and he grows up in the very heart of the Upanishadic country genuinely reflecting its tensions and concerns.

> In the shade of the house, in the sunshine on the riverbank by the boats, in the shade of the sallow wood and the fig tree, Siddhartha, the handsome Brahmin's son grew up with his friend Govinda. (3)

This stark opening line sweeps the reader into the narrative, and sets the tone for what is to follow. In fact, the elemental quality of the sunshine and the shadow by the riverbank would contribute to the fable-like structure of the tale. The river is as significant as the protagonist of the narrative. Is it a mere tale? Or is it the written records of one who has suffered and thought profoundly in isolation and grief at the human condition? This fable-like quality of the narrative would add to its distinctness and genuineness alike. Hesse has divided his work into two parts: the first contains four chapters from *The Brahmin's Son* to the *Awakening*; and the second from *Kamala* to *Govinda* contains eight chapters. The pains and pleasures of growing up and growing apart, of knowledge and experience, of love and wisdom—all these are traced in the tale of Siddhhartha. He is the main focus of attention and all others including the river, the land, sky, and landscape are seen from his perspective, although the narrative itself is in the third person bestowing a suitable distance to the agonies and anxieties of the protagonist, till the real encounter with the river comes to be. Siddhartha is deeply meditative and completely committed to his quest. Along with his trusted friend Govinda, he leaves his father and joins the wandering Samanas in search of abiding knowledge. There is so much wandering in the tale, a reflection of one of the most enduring issues of the twentieth century—the unending search for the common ground of self and soul, the indomitable quest for the unified experience of the body and the spirit, the unified experience of the inner and outer nature. Siddhartha however, finds the heart returning forever to cycles of existential experience:

> But although the paths took him away from the Self, in the end they always led back to it. Although Siddhartha fled from the Self a thousand times, dwelt in nothing, dwelt in animal and stone, the return was inevitable; the hour was inevitable when he would again find himself in sunshine or in moonlight, in shadow or in rain, and was again Self and Siddhartha, again felt the torment of the onerous life cycle.(13)

The chapter entitled *Gotama*[**] is significant because of the brief interlude of Siddhartha's encounter with Gotama the Buddha:

> He wore his gown and walked along exactly like the other monks, but his face and his step, his peaceful downward-hanging hand, and every finger of his hand spoke of peace, spoke of completeness, sought nothing, imitated nothing, reflected a continual quiet, an unfading light, an invulnerable peace.(23)

It is significant how Siddhartha observes the Buddha's body attentively: "and it seemed to him that in every joint of every finger of his hand there was knowledge; they spoke, breathed, radiated truth."(23) And much later he would tell Kamala: "Among all the wise men, of whom I knew so many, there was one who was perfect... .He is Gotama.(58)"

However, Siddhartha's quest does not end with the encounter with the Illustrious One which only gives back to him his own self. While Govinda chooses to follow the Buddha's path, Siddhartha sets forth again. The thinker in him awakens and he muses:

> Yes, he thought, breathing deeply, I will no longer try to escape from Siddhartha. I will no longer devote my thoughts to Atman and the sorrows of the world. I will no longer mutilate and destroy myself in order to find a secret behind the ruins. I will no longer study Yoga-Veda, Atharva-Veda, or asceticism, or any other teachings. I will learn from myself, be my own pupil; I will learn from myself the secret of Siddhartha.(31)

His recognition that meaning and reality were not hidden somewhere behind things but they were those things themselves is the beginning of another cycle of learning. The second part of the narrative is the story of this discovery when Siddhartha moves from one extreme to another. With Kamala, who represents the sensual pleasures and limits of the corporeal being, with Kamaswamy, who represents the knowledge of the world of business transactions, and much later with Vasudeva, the ferryman who represents the third bank of the river of learning, Siddhartha learns to share many aspects of existential truths and realizes the multiple dimensions of worldly existence. And finally after he manages to extricate himself from all the entanglements of life and its innumerable demands he finds himself beside the river. This is the very same river across which the ferryman had once taken him when he was still a young man. Now, beside himself with the anguish and suffering of the body and soul Siddhartha wrestles with the terrible emptiness within him contemplating suicide. But the reverberations of the *pranava mantra*, Om, bring him back into his higher senses, and here he sinks into a profound sleep. Govinda encounters him sleeping by the riverside and the two friends once again debate and recognize the crossings in their life's individual paths. After he watches Govinda depart Siddhartha continues in his struggle to come to terms with his own experience. "Now he understood it and realized that the inward voice had been right, that no teacher could have brought him salvation. That was why he had to undergo those horrible years, suffer nausea, learn the lesson of the madness of an empty futile life till the end, till he reached bitter despair, so that Siddhartha the pleasure monger and Siddhartha the man of property could die. He had died and a new Siddhartha had awakened from his sleep." (79)

The river flows into the narrative once again. Siddhartha's wanderings have the same familiarity like the river's meanderings. They are all the same and yet not the same at any time. But beside the river his meditations achieve a greater degree of clarity. "Will you take me across?" he asks the ferryman. And in Vasudeva he finds a passive listener to whom he can narrate his life's experiences. Vasudeva draws his attention to the river: "the river has spoken to you. It is friendly towards you; it speaks to you." (84) Siddhartha stays with the ferryman and "learned more from the river than Vasudeva could teach him. He learns from it continually. Above all, he learns from it how to listen, to listen with a heart, with a waiting, open soul, without passion, without desire, without judgement, without opinions." (85) This dialogue with the river takes up a supreme space in the novel. The river talks to Siddhartha and he learns to listen. The river stands for articulation and silence, mobility and stillness, continuity, change and presence, at the same time. If the act of crossing the river earlier was symbolic of a change of environment and transition from one geographical space to another, the second crossing is one of transformation and compassion. Vasudeva tells him:

> ...the river is everywhere at the same time, at the source and at the mouth, at the waterfall, at the ferry, at the current, in the ocean and in the mountains, everywhere, and...the present only exists for it, not the shadow of the past, nor the shadow of the future...(86)

This is in fact what Siddhartha realizes. He learns to abide by the river and listen to its murmurings. It is while he lives there with the ferryman that he comes to know about the passing of the Buddha, and compassion and good will permeate his very being. It is here that he comes across Kamala, the courtesan once again, and she dies beside him smitten by a snake. It is then that his son is joined with him once again. However, as he was soon to find out, there was a world of difference between himself and his young son—their pursuits were so different. And soon enough the boy picks up a quarrel with him and rows off in their boat. Siddhartha is broken hearted and behaves like a householder setting off in search of his son. However a sense of purposelessness in his acts and deeds soon comes over him and his higher sense of listening to the river brings back his questing self all over. Nevertheless he does not despise the simple commitments and simpler bondages that tie human beings to this world; they are so inevitable and necessary in the vital actions of the indestructible *Brahman*. This wisdom mellows in Siddhartha: the consciousness and unity of all life, a holistic sense of unending compassion. This wisdom is unique and incommunicable. This is the result of an existential and experiential encounter. The rest of the novel is the passing of Siddhartha, his final adieu to this world and his lifelong companion Govinda, who is his own surrogate self. This double vision is not unique to this work of Hermann Hesse alone, it is an underlying concern in all his works alike. Everything is valuable and holy, nothing is insignificant; this is the inscrutable philosophy of existential truth that Siddhartha embodies in the final analysis. The sacred is never on the other side of the self, the *sacred is the self* nor are we out of it!

Siddhartha is a magnificent work of search, vision, and understanding. There is a rebellion at the heart of spirituality that never accepts the commonplace and ordinary as constituting the entirety of existence; this is brought home in the guise of Siddhartha who is even not content with being a follower of the Buddha, the Compassionate one, but ceases not till the truth of experience is brought home through his own lived experience. This sense of individual enterprise is perhaps the outcome of the individualistic humanism of Hesse himself in his encounter with the values embodied by the *other side of the river*. In reaching toward the *third bank of the river*—a logic that defies reason—the novel builds itself on a texture of elemental harmony that defies historicity.

Above all, *Siddhartha* continues to be a classic work of sympathetic understanding—of humanism and spirituality at the same time. Living is an experience that is inexhaustible, and the quest for the spiritual at the heart of being only reinforces and strengthens this inexhaustibility.

❖❖❖

In our present day world driven by market economy where all values are reduced to their commodity status and prices, where ignorance and imbecility coupled with collective laziness imposed by the hegemonial order of technology, together, but serves to only reduce and mock at the very concept of spiritual and the questing mind, works like *Siddhartha* might sound a bit odd and out-of-place. After all, in which century, in which times, in which cultures, have the concern and commitment of sages and saints, mystics and prophets, philosophers and poets, ever sounded sensible and sensitive, relevant and meaningful to the common folk who appear to be concerned only with the daily struggle of existence? Those values for which Socrates died and the Buddha renounced his all and everything are not so simple enough to be internalized by the insensitive and unconcerned. It requires a wideness of concern and a profundity of sense and a will to sympathise, to understand and *to listen*. The mystical is of course beyond any attempt at historicisation, and hence transcends logic and reason.

> When Siddhartha listened attentively to this river, to this song of a thousand voices; when he did not listen to the sorrow or laughter, when he did not bind his soul to any one particular voice and absorb it in his Self, but heard them all, the whole, the unity; then the great song of a thousand voices consisted of one word: Om—perfection. (107)

Passion and compassion are at the very core of Siddhartha. In a perpetual search for meaning and spiritual realization

Siddhartha wanders from place to place, and his quest sensitizes him to the various potentials of human's being. There is a *fullness* at the heart of the novel that seeks for a complete satiation of all sensual possibilities and a fuller realization of non-human nature, leading toward a transcendent completeness. It is not that Siddhartha stands apart from us in time, place, language, history and culture. He is like us who are encapsulated in a *globalising* present. In the consumer-centred transnational market economy of the present *the pleasure principle* looms dominant.

> Pleasure seeking is a duty since the consumption of goods and services becomes the structural basis of Western (read all civil) societies. Social integration thus takes place less through the principles of normalization, confinement and disciplinary power, as described by Foucault. Instead it takes place through the "seduction" of the marketplace, through the mix of feeling and emotions generated by seeing, holding, hearing, testing, smelling, and moving through the extraordinary array of goods and services, places and environments that characterize contemporary consumerism organized round a particular "culture of nature". (*Contested Natures*, p 24)

In a strange manner the experience of *Siddhartha* is no different. He is caught in the miraculous world of worldly existence—*samsara*—which ultimately reveals itself to be mere tinsel and a naught. However until this realisation is borne home in him—not merely as an idea but experientially—he has to seek for fulfillment without seeing, holding, hearing, testing, smelling, having and possessing. Kamala, the pleasure grove, the world of Kamaswami, and all that it holds had to be experienced for what it was—mere transitory forms. We read in the chapter entitled By the River, which also structurally occurs in the middle of the book accentuating the image of the river around which the narrative turns:

> Smiling he listened to his stomach, listened thankfully to a humming bee. Happily he looked into the flowing river. Never had a river attracted him as much as this one.

> Never had he found the voice and appearance of flowing water so beautiful. It seemed to him as if the river had something special to tell him, something which he did not know, something which still awaited him. Siddhartha had wanted to drown himself in this river; the old, tired, despairing Siddhartha was today drowned in it. The new Siddhartha felt a deep love for this flowing water and decided that he would not leave it again so quickly. (79-80)

An aesthetic experience of distancing as well as engagement is what Siddhartha undergoes. And this happens so naturally. There is nothing miraculous about this experience—it is not extraordinary or out-of-this-world. It does not deny the sensual experience or ecstasy of this world, but, on the other hand, extols the power and force of the life forms that constitute outer nature. From this point onwards Sidhhartha undergoes deep feelings of compassion and primal sympathy for all life forms, human and non-human. This aesthetico-spiritual experience is a transforming force and an integrating one. This is the beginning of eco-dharma, the ethical vision of a holistic nature.

The very next lines that occur in the novel are:

> I will remain by this river, thought Siddhartha. It is the same river which I crossed on my way to the town. A friendly ferryman took me across. I will go to him. My path once led from this hut to a new life which is now old and dead. May my present path, my new life, start from there. (81).

The river is within him now. He learns to see and sense and listen. The inner and outer nature have become one—a holistic experience of being. Perhaps when one extols the virtues of eco-dharma and eco-spiritual values, one needs to recognize the critical eye that would call spirituality and all its inner experiences to task for being essentially subjective, internal, depending in all, only on the inner inexplicable experience of an individual. Of course, the spiritual is primarily an experience that affects the inner individual responding with his/her inner sense, the *antahkaranas*. But this is not to be negated, nor countered as of little value by the collective social consciousness that a socially self-aware individual would perhaps, internalize. Of course, in order to identify with the *wretched of the earth* the social aspect of the individual awareness needs to be awakened. Similarly in order to be sensitized to the desecration of the earth and all its denizens spiritually or otherwise the inner sensitive spiritual being needs to be awakened. After all, it is only by the sensitization of the self-reflexive being that the individual finally becomes one with the collectivity we term as society. All commitment to the collective begins by being self-aware, as individual, the citizen, or the corporate term now in use, the stake-holder! The ancient Indian idea of state incorporated the specific space for both the *vyashti* (the individual) and *samashti* (the collective). The self-aware individual seeks for a holistic experience that completes his/her internal being in the collective becoming of the spiritual! Eco-spirituality is thus a reordering of the individual's process of completion of his/her selfhood—in society and in the Spirit at the same time. This is not to reinterpret ecologically sensitive critical spirit as one that is formless, Protean, or amorphous, but only to reinvest in the values of the *polymorphic* and the *heterogeneous*. Hesse's *Siddhartha* hastens us into the discourse of the holistic through the dialogues of the self and soul, of both the individual and the society at large.

Eliot Deutsch, in a perspicacious essay reasons thus:

> Nature is in "fact" not "value." Nature can become value laden... only from the spiritual perspective which sees nature either as a manifestation of spiritual being or, Advaitic-wise, as an appearance grounded in a spiritual reality. [2]

This is the elemental wisdom that Siddhartha unfolds deep within through his all-transforming spiritual insight. "Are you going into the woods?" Siddhartha asked the radiant Vasudeva. "Yes, I am going into the woods: I am going into the unity of all things," says Vasudeva, radiant. And so he goes away. Siddhartha watches him. With great joy and gravity, he watches him, sees his steps full of peace his face glowing, his form full of light (108). Wisdom, as Siddhartha realizes soon enough, is not communicable; it needs to be felt and experienced. And the world that incorporates both the individual and the society, in the final analysis, is not an illusion, it is also a felt experience. When Govinda asks him, "your stone, your tree, are they real?" Siddhartha replies with the sedate calm of a self-realised mystic:

> If they are illusion, then I also am illusion, and so they are always of the same nature as myself. It is that which makes them so loveable and venerable. That is why I can love them. And here is a doctrine at which you will laugh. It seems to me Govinda, that love is the most important thing in the world. It may be important to great thinkers to examine the world, to explain and despise it. But I think it is only important to love the world, not to despise it, not for us to hate each other, but to be able to regard the world and ourselves and all beings with love, admiration and respect. (115)

This is the ultimate expression of an eco-spirtual vision. The ability to reason and love at the same time, this is the point where the intellect, the intuition and the senses all act together. A self-realization of eco-dharma would mean being able to exist as an independent being and also being able to love all and everything simultaneously as socially and spiritually responsible being. Siddhartha's experience is no different. This is the reinstating of the sacred at the heart of all being, natural and unnatural, human and non-human, the all-transforming vision of being and becoming sacredly. For our world of the present which is devoid of any semblance of the sacred where earth and sky and space are looked upon as mere resource for human's being Sidhhartha hastens an eco-spiritual, value-ridden, meaningful universe, that reverberates with the sacred and the beautiful, the ananda of being, like the experience in Basho's haiku I had provided as an epigraph to this text: "Ah, how glorious! Green leaves, young leaves/Glittering in the sunlight."

Notes and References

1. *Siddhartha*, translated by Hilda Rosner, Madras: Macmillan, 1973. All citations that follow are from this edition. (Page numbers are provided within brackets). Hesse published *Siddhartha* in 1922. The first *British Commonwealth* edition was released in 1954, by Peter Owen Ltd. The Macmillan India edition was under arrangement with Pan Books Ltd., London, who brought out the *Picador edition* in 1973
2. "A Metaphysical Grounding for Natural Reverence: East-West," *in Callicott,J Baird, and Roger T Ames, Ed. Nature in Asian Traditions of Thought: Essays in Environmental Philosophy. 1989;Indian reprint, Delhi: Sri Satguru Publications, 1991. p 264.*

* Usually spelt *Gautama*

** As part of a project of the *Hermann Hesse Society of India* to make available *Siddhartha* in many Indian languages, I was invited to contribute a general introduction common to all the editions. This entailed a new re-reading of *Siddhartha* (quite distinct from the modernist space) and this paper is a natural extension of that experience.

3

Ecomorphic Horizons: From the Dark Ages to the Romantic Big Bang of Culture and Beyond[1]

— Mihai A. Stroe

Introducing such a vast notion as that of "ecological man" must constitute in itself an endeavour to discover networks of moments in history—in our case, the history primarily of science and literature, but also of religion—which more or less intensely and more or less continuously communicate with each other, thus constituting a matrix of thought of different degrees and inflexions pointing to a definitory knowledge or science of the relationship between man and nature. We are thus mainly interested in the moments in history when perception of man as a being inside the cosmos shifts between different reference points that are themselves fleeting points, fluid frameworks of self-perception. To attach the notion of "ecological" to man must thus be in itself a definitory act of science and philosophy at the same time.

The evolution of the concept of ecological man in history is otherwise like the history of an iceberg; for us to speak today of an "ecological man" we must already conceptually be standing on the mulscular shoulders of intellectual, philosophical, scientific, religious and artistic giants who laid layer after layer of cognitive energy that in the end ossify and become stable—dynamic structures having in themselves an embedded instability: out of the tension between this stability in instability and instability in stability a pendular vital force of thought and cognition was and is perpetually born, a pendulum of tendencies that fundamentally determine the behaviour of man in the universe. We shall call this basic process the *Pendulum of History*, as a concept derived from the ideas of Werner Heisenberg and Arthur Koestler, forming a kind of universal cognitive matrix of thought by which man defines himself in relation to everything.

Thus, when speaking of "ecological man" we speak of a way of perception, a way of seeing our place in history and the cosmos. It seems that this perception is one of the most fundamental driving forces: the force of self-definition in relation to everything else, out of which all behavioural consequences are derived. In other words, believing ourselves to be "ecological men" we in fact determine our behaviour to fit that state of affairs not only essentially (in our interior life), but also in our exterior life. From this play of the dialectics or duality of cognitive perception man learns to define his place in both history and the cosmos, even if while so doing man can undergo dramatic transformations. In this sense, a new discovery in science can overturn what long centuries had hoped to develop towards a final, inevitable and irrefutable definion. From this veritable dance of cognitive energies man emerges as a complex being concerned with his ontological status.

It seems that the evolution of man from the Middle Ages to Romanticism and beyond to the postmodern and post-postmodern age results in the increasingly clearer definition of man as an ecosystem in itself, evolving from an infra-natural level/dimension to a natural and finally to a supernatural dimension. In other words, man rises in the second millenium after Christ gradually from the status of a being crushed under the weight of the whole of nature, like a defeated giant living underground. For this reason, the Middle or Dark Ages could well be called an "atlantean age", in the sense that man resembled in many ways the old mythical figure of Atlas, bearing the entire burden of the cosmos. The Middle Ages is an Atlantean age because man is morally burdened under the dawning realization of the weight of his deep fall from Eden, Nature being drawn in the process alongside with him. Because

Nature herself was involved in the deep fall, medieval man feels the burden also as something physical, expressed in the weight of religious interdictions of all kinds and the institution of the Inquisition. This age best typifies man's paradoxical fall into the depths of Nature, who follows man in his Fall, according to Christian tradition. As the poem *Everyman* proclaimed, nature is now associated with vanity: "All earthly things is but vanity", an echo of *Ecclesiastes* 12:8. But in the Middle Ages Nature gradually also begins to be associated with the idea of an "Earthly Paradise", as in *The Romance of the Rose*, or in *Mandeville's Travels*, where the "Paradise Terrestre" is mentioned as being "the highest place of earth that is in all the world, and it is so high that it toucheth nigh to the circle of the moon." This image was based on the Ptolemaic system, with the earth at the centre, the lowest sphere being that of the moon, under which all was governed by mutability. This "Paradise Terrestre" as the garden of Eden was unreachable, because Nature herself blocked the access of man by interposing "wild beasts" in the deserts, "high mountains" and "great huge rocks", "great tempest", etc. Only the "special grace of God" could open access for mortal man to this terrestrial paradise. This view of the terrestrial paradise later opened the perspective for a romantic like John Clare, who came to see in Nature something "more than eden"[2] (cf. Decay, in Clare, 2007: 210-211), like an excess of the supernatural inside the natural, this excess being needed as a counterpoise to the vast burden of darkness: only through such romantic excesses were an escape out of darkness and the leap to the transcendent realms of spirit in the first place possible.

An important predecessor to Clare's ecomorphic view seems to have been Andrew Marvell (1621-1678), who in his poem *The Garden* anticipated an eco-romantic worldview based on a Goethean mighty analogy between micro- and macrocosm: "The mind, that ocean where each kind/Does straight its own resemblance find;/Yet it creates, transcending these,/Far other worlds and other seas,/Annihilating all that's made/To a green thought in a green shade." In other words, the transcendental Nature created by the imagination is far more beautiful compared to physical fallen Nature. Moreover, as later William Blake and other romantics emphasized the view on the *horror vacui* in the guise of the romantic principle of infinite abundance, Marvell also emphasized that nature is a lover of abundance, as in *An Horatian Ode upon Cromwell's Return from Ireland,* 41: "Nature, that hateth emptiness." As Harold Bloom has shown, the major concern of the romantics was precisely this quest of the poetic imagination aimed at reaching not necessarily just a unity between man and nature, although this was implicit in most of the cases, but a surpassing of nature followed by an embracing of the spirit's infinite transcendence.

During the Middle Ages man reached to Nature, thus becoming buried deep inside the materiality of the Natural—like Blake's Urizen later, buried deep inside the jungle or the forest, but with no access to a paradisian dimension, excepting some isolated cases such as of certain saints or alchemists / shamans. Many gnostic myths deal with this idea, gnosis going even so far as to state that Christ himself fell and forgot his own identity under the weight of matter and its darkness. St. Augustine, however, had warned as early as the fifth century AD that in fact the reason for man's evil character is not matter per se, and so consequently not Nature. Evil was somehow generated by man himself, by virtue of his free choice, from the inner space of his own will, which being free was able generate it. The Middle Ages are so to speak "dark" because now man's capacity to "see" inside darkness in all its amplitude is sharpened. Once seen, darkness becomes a visible enemy against which now man begins to form weapons for defence. Bunyan's Christian, although a Renaissance figure (1628-1688), is still such a case of deep paralysis in front of the realization of one's sinful condition. Is still such a case of deep paralysis in front of the realization of one's sinful condition. Only the glimpse of the divine and of human destiny makes it possible for Christian to begin his spiritual rise from the Dark Ages and into the Renaissance of man, a rise that had already been announced by the birth of perspectivism in the last phase of the Middle Ages: with the gain of perspectivism—such as in Alhazen first (the 11th century), then in Albrecht Dürer, who travelled to Italy in 1506 precisely to learn about "the secret art of perspective", and in others, man gains a powerful cognitive vision of the physical world: the Renaissance is a cognitive age, when man comes out of the depths of Nature and its darkness and leaves the burden of sinfulness on Nature's titanic shoulders herself, or, as in Christian's case, on Christ's shoulders.

The fascination with perspective entered art, especially in the north of Italy (in Florence and Venice), in the fifteenth century, leading to the emergence of the school of Perspectivi, associated with such artists as Lorenzo Ghiberti, Fillipo Brunelleschi, Carpacio, etc. this being a school of thought aimed at creating a sense of life, of movement in a space with three dimensions. Alhazen had clarified the deep nature of perspectivism; namely, the cone of rays coming from the shape of the object grows narrower as we move away from it, so as it grows larger as we move towards it (Bronowski: 112). Unlike other scientists, Roger Bacon was among the ones interested in this phenomenon. The shift was profound, from a God's eye view, such as in the paintings of 1350, which attempt to capture the ways things are, as a map of eternal truth—this being the case of the Plotinian paradigm—to a view of how things look, this being the new da Vincian paradigm. Thus, the perspective painter leaps away from any absolute and abstract view, and what he aspires to capture is not a place, but a fleeting moment: a point of view in time more than in space, as Jakob Bronowski keenly observed.

In the Middle Ages, therefore, God had been central and so the ruling paradigm was theomorphic, with man crushed under the burden of matter. During the Renaissance, on the other hand, the ruling paradigm becomes anthropomorphic and chronocentric: man and time become central. Now Nature becomes the "writer of beauty", as Sir Philip Sidney (1554-1586) put it in *Astrophel and Stella*, 4: "in Stella's face I read/What love and beauty be; then all my deed/But copying is, what, in her, Nature writes." To some degree, man's rise in eminence occurs in any new rebirth paradigm. The Renaissance is thus governed by natural man, who had sporadically made his presence felt even during the Middle Ages in such personalities as Roger Bacon, Arnold of Villanova, Raimundus Lullus, etc. all of them prominent figures of scholars—alchemists who were trying to gain universal knowledge one way or another, and form universal systems of thought to be used for man's benefit in life.

Roger Bacon aimed in this sense at total knowledge which was to purify even Christendom itself, and so transform and convert the world. In fact, it should be reminded that Roger Bacon was the one to have suggested to the Pope, his patron, Guido de Foulques (later Clement IV), that it was desirable to have the whole world surveyed, so much so that Bacon drew a map of the world showing the sea route westward from Spain to the Indies, and this suggestion came down to Columbus, via Peter of Ailly, who thus set sail to conquest the New World, at the dawn of the Renaissance. To Roger Bacon, however, Nature was an instrument of the divine plan of operation which is discernible in the history of the world. As a consequence, he abandoned the idea of apocalyptic prophecy, and so adopted a "scientific" astrology, as most Franciscan Spirituals did. Roger Bacon was convinced of the imminence of the new and last age of spirit which was to be introduced by "spiritual man", i.e. the enquirer into nature, the pure scientist, who was to bring enlightenment and universal peace to the Church and the world. At this point, another important vast shift took place announcing the Renaissance: the shift from the notion that the agents of "God's grace" are emperors, kings, popes, rulers, all of them bearers of salvation, to the notion that "God's grace" and bearer of salvation was machines, gimmicks invented by man through God's intercession. It should, however, be also reminded that at this time scientific experiment was deeply intermigled in magic and sorcery, of the kind announced by Faustian images in literature. The same is true about da Vinci's own time. In fact, as Friedrich Heer justly observed, at any time between the Middle Ages and the nineteenth century there were princes all over Europe supporting natural "scientists" in the hope of producing gold, powerful machines of war, or the elixir of life. In other words, natural science, which in parallel developed into a metaphysics of nature known as *Naturphilosophie*, was at all times also associated with alchemical thinking, it only gradually and partially being separated from it in the birth of modern chemistry.

Roger Bacon's disciple, Arnold of Villanova (born 1238), typifies in this sense the alchemical search, in this case for a medicine which should cure all ills. For Arnold of Villanova, natural science was a tool, as was for most of the Spiritual Franciscans, to be used in remaking the world. This stance comes rather close to the new views of the holistic paradigms with their aim at total reformation of the world, with a view to man's gaining infinite knowledge, freedom, and life. Also, we should mention that, as Fr. Heer emphasized, during the thirteenth and fourteenth centuries, Petrarch and the humanists launched a relentless attack on what they called "godless physicians", referring mainly to the Averroists of Padua. The mistrust thus generated went on for the next 500 years, and, it would appear, is still present now in humanists, artists, creative individualists, to the extent that doctors are seen as being aggressors against man, i.e., against the holiest of things on earth. But even in these circumstances, the natural sciences prospered in the margins of Europe, in marginal groups of society, being practiced by people working outside the university, namely, by Jews, Muslims, and Spiritual Franciscans.

Ramon Lull (1235-1315), on the other hand, unified in his work and life the vision of medieval science and theology, of medieval Church and State. In his view, the conversion of Christianity was *true enlightenment*, so much so that faith and reason, mystery and rationality were indissolubly wed. Lull had the vision of an all-embracing logic, his *Ars Magna*, aimed at reducing all existing languages to one. This was his dream of universal knowledge, expressed in a computing engine, which connected the basic principles, or "Ground-Words" as he called them, of all knowledge by a mechanism made up of concentric circles segmented by radii and by geometrical symbols. This was a sort of cybernetic machine, a miracle machine, prepared to solve every problem, every science, even faith itself, of the kind that Swift was later to mock in his *Gulliver's Travels*. Leibniz, however, did not mock at Lull's vision, since it came too close to his own dream of finding a universal, scientifically viable language, his *Mathesis Universalis*, which was to enlighten all men impartially, this being the prerequisite of universal peace, the kind of peace now searched by virtually all deep ecological, ecosophical movements around the globe.

In this sense, Fr. Heer is right in observing that Lull's idea anticipated the great universal formulae of Einstein and Heisenberg, which provided us with the mathematical keys to the problems of matter, light, energy, and the laws of the cosmos. In this sense, we should perhaps add that the formalization of science was anticipated also by a lesser known medieval figure, namely, Albert of Saxony (1316-1390), born one year after Lull's death, who is known to have been among the first to attempt to find out mathematical formulas by which to express laws of nature, thus foreshadowing modern physics. Also to be noted is the fact that Ramon Lull influenced through his thought great personalities such as Nicholas of Cusa, Pico dela Mirandola, Giordano Bruno, Gassendi, Leibniz, etc. his ideals being also present at least partially in the romantic revolution, in radical cries such as Blake's in *Jerusalem*:

> "I must create a system or be enslav'd by another man's."

In the Renaissance man thus tends to more sharply realize that he possesses two priceless gifts: *The Book of Nature* and *The Book of Scripture*, as codes for the decoding of life's meaning and mystery. In this sense, in 1481 the German priest Konrad von Megenberg published his *The Book of Nature* which is the first printed work containing images of animals. The idea of a book of nature was mentioned for instance by Sir Philip Sidney (1554-1586) in his *Astrophel and Stella*, 71,[3] but in his *Defence of Poesie* he considered nature's world as being "brazen"—therefore mutable, while that of poetry—"golden", therefore immutable, eternal.

The increasing interest in Nature as deep mystery can be observed in the creation of associations such as *Academia Secretorum Naturae The Academy of Nature's Secrets*, founded in Italy in 1560 by the physicist Giambattista della Porta, and being aimed at the exhange of scientific information. This interest can also be discerned in the increasing number of publications dealing with "fossils" in the following period, a term popularized by the German metallurgist Georgius Agricola (Georg Bauer) in his *De Natura Fossilium*, published in 1546, a term by which he understood everything that is dug out of the earth (Lat. *fodere* = to dig), including rocks having the shape of bones and shells, and even amber containing insects and spiders.[4] Later, in 1669, fossils were shown by the Danish geologist Nicolaus Steno to be the petrified remains of ancient beings, his view being eventually accepted. (The English naturalist John Ray was among the first to believe in this idea, as early as 1691). The first published images of fossils appeared as early as 1565 in *De Rerum Fossilium/On Things Dug out of the Earth* by the Swiss naturalist Konrad von Gesner, who thus brought to light the oldest known catalogue of natural curiosities belonging to Johann Kentmann (1518-1574), who in this way contributed to the emergence of the first so-called "curiosity cabinets" such as that owned by himself, exhibiting "fossils" in Agricola's sense, and which became more and more fashionable in Europe in the next two centuries: for example, the "Metalotheque" of Michele Mercati (1541-1593), whose description was published only in 1717; the "Musaeum metallicum" of Ulisse Aldrovandi (1522-1605), whose catalogue was published in 1648; the famous curiosity cabinet of Francesco Calzeolari (1522-1609), described in *Musaeum Calceolarianum Veronese* by Benedicto Ceruti and Andrea Chiocco published in 1622.

On the other hand, Konrad von Gesner's importance surpasses the fact that he was the first popularizer of fossil images. He was the godson of the Protestant reformer Ulrich Zwingli, and is considered to be the founder of zoology for his monumental work in five volumes *Historia animalium*, published between 1551 and 1558. In fact, Gesner became known also as "the Father of Bibliography" for his other monumental work, *Bibliotheca universalis* (cf. J. Christian Bay, 1916), which is the first bibliography containing about 1,800 authors, alphabetically listed, with evaluations on each author's works. This was a first vast signal addressed to the public on the existing sources of knowledge: in the subtext, this signal might have pointed also to the danger that these sources of knowledge ran the risk of becoming themselves cultural fossils. This systematization by Gesner became a powerful stimulus for encyclopedic learning in the Renaissance and for what would come to be the ideal of the *homo universalis* as a *summum* of the human, integrating natural science, religion and art.

Moreover, it is relevant to point out that Leonardo da Vinci (1452-1519) seems to have been among the first to meditate on the nature of fossils in rocks, rejecting the idea that they were generated by astral influence or as a consequence of the Flood (the hypothesis supported then by the fathers of the church such as Luther; according to the latter, fossils were witnesses of the Flood). By observing the morphology and locations of fossil shells and by comparing them with the modes of life of living shells, da Vinci concluded that the fossils were traces of once living creatures. Interest in fossil formations, however, had occurred earlier in Europe, the evidence thereof being preserved in such writings as *Speculum Mundi* by Vincent de Beauvais, a prototypical encyclopedic work, written ca 1250, showing special concern for rocks; or as *De Mineralibus* by Albertus Magnus (1193-1280), written ca 1260 and printed only in 1476, in which he speaks about "the stones having animal forms in their inside or outside parts", which had been generated (as Avicenna, too, had explained) as the petrification of animals under the mysterious action of an emanation of petrifying force.(Book I, chap. VIII).

The interest in fossils and the natural was combined and thus inhanced by the rise in intererst for the ruins of former civilizations. The birth of archeology is thus considered to have been inaugurated in 1592 by the Italian engineer Domenico Fontana, through his sensational discovery of the ruins of the ancient cities of Pompeii and Herculaneum—buried by the volcanic eruption of the Vesuvius in AD 79. Time thus becomes a powerful and mysterious object of philosophic wonder and curiosity, primarily defined as an indomitable force to transform everything into ruin, as Sir Thomas Browne put it in his *Hydriotaphia*, or *Urn-Burial* (1658), V: "Time which antiquates antiquities, and hath an art to make dust of all things."

It would appear that the period of the Renaissance and Enlightenment mediated a vast accumulation of puzzles and clues that later constituted the huge tension having triggered the Romantic cultural Big Bang. In fact, contemporary holistic thinking, with deep roots in Romanticism, such as expressed in Nerval's view of the "vast conspiracy of all animate beings to re-establish the world in its first harmony", may have evolved precisely through the rise in interest regarding the mysterious quality of time—with all its visible and invisible dimensions—as both a distructive and creative power, and as a paradoxical deeply mysterious medium into which the cosmos and its dynamics is profoundly steeped, such as for instance Giovanni Battista Piranesi's famous work, the *Carceri d'Invenzione*, suggested: a temporal labyrinth in which human imagination is trapped, but also entranced, as Sir Thomas Browne sharply noted, again in his *Hydriotaphia*: "there is no antidote against the opium of time."

The Romantics were the first to have set as their ultimate purpose escaping the natural labyrinth of time and conquering the infinite labyrinth of spirit, even though the travel through the abundant temporal depths was part of the joy of spiritual adventure, a path cut through deep time as their imaginative *terra incognita*, their natural labyrinth that, once explored, was supposed to be mapped for the sake of future generations to be involved in the vision quest.

A strong impetus towards holistic thinking or integrativism might well have been given in 1651 also by William Harvey through his hypothesis that all beings come from an egg (the *ex ovo omnia* theory), this being tantamaount to postulating a universal principle of life's origin. Time was thus essential not only for a scientific understanding of life, but also for a religious and generally cultural understanding. Perhaps a culmination of this interest for the workings of time is constituted by two episodes, having happened in Europe of the 1790s. W.J.T. Mitchell told the story of this in his remarkable article entitled "Romanticism and the Life of Things: Fossils, Totems, and Images" (2001).

One of the two episodes took place in Paris, in 1795, when George Cuvier reconstructed a mammoth (discovered alongside the rivers in Siberia), wishing to demonstrate his theory of "fossils", which became an instant international sensation in intellectual circles: his program was aimed at transforming the fossils—then still seen as monsters—into what they really were, namely "traces of extinct life" (as the Danish geologist Nicolaus Steno had suggested) and proof for the fact that the earth had gone through a series of catastrophic geological revolutions in its history that wiped out various life forms, these cataclysmic phenomena being preserved in the deep structure of the earth. (A first work in which the geological strata are classified in terms of the characteristic forms of fossils was published by William Smith, the "Founder of Stratigraphy", in 1815: *A Delineation of the Strata of England and Wales, with Part of Scotland*).

The second episode, rather obscure, took place in London, in 1790: the story of how the word "totem" entered science via the memoir of an obscure Englishman, John Long, who had worked as a fur trader among five nations of Canadian Indians and had been initiated as a Chippewa warrior while serving as a military advisor to these tribes who were warring as allies of the British army against the American colonies. John Long returned to England carrying on his chest a tattoo in the shape of a beaver which he called "totem"—an Ojibway word,[5] meaning "he is a relative of mine" and, as Claude Lévi-Strauss suggested in *Totemism* (1963), associated with the idea of animal, vegetable or mineral "tutelary spirits", and so with the idea of destiny, identity, and community. John Long's memoir was favourably reviewed in 1792 ("Voyages and Travels of an Indian Interpreter and Trader", in *Monthly Review*, June, 1792: 129-137), and was soon translated into German (1791, 1792) and French (1794, 1810). The notion of "totem" thus unpredictably spread in the academic community to become a foundation of anthropology, comparative religion, and even sociology and psychology. (See Freud's *Totem und Tabu*, 1913; see also Emile Durkheim, *The Elementary Forms of Religious Life*, 1912, in which the author mentions John Long's book and the idea that totemism was known—for more than half a century after the publication of Long's volume and because of it—as being an exclusively American institution. Cf. the chapter on "Totemism as Elementary Religion", in Durkheim, 2008, I, 4: 77). Mitchell's hypothesis is that both "fossil and totem are windows into deep time and dream-time respectively", as well as windows into "the deep past and the deep present of physical things from the different angles of nature and culture." Indeed, "the deep past" and "the deep present" were two phenomena which fascinated the romantic mind, and in this sense Mitchell poses the following intriguing question: was the French Revolution a product of the new sense of natural history as uncovered by George Cuvier by his theory that fossils are a trace of a life-form that was erased from the face of the Earth via geological revolutions, or is it that Cuvier's radical theory is just an echo of the French Revolution itself. Mitchell answered indirectly by emphasizing that, whichever the case might be, the notion of fossil became a metaphor for both natural and human history, and especially for the "human relics" left behind by the French Revolution. In other words, "fossilism" becomes a way of "revolutionizing natural history", while at the same time "naturalizing revolutionary human history" (Mitchell: 176). As a consequence, Mitchell suggested that Paul de Man's theory about the romantic image should be revised in the sense of considering the latter as being "a composite of two concrete concepts, the fossil and the totem", equivalent to the energies of petrification (fixity, the finite, matter: fossil) and animation (mobility, the infinite, spirit: totem). Totemism points to the longing for an intimate relation with nature and all its tutelary spirits—expressed in Wordsworth's daffodils, Shelley's west wind, Coleridge's albatross, Blake's sick rose, Novalis's blue flower, etc.—that are not to be killed, and when they are tragically killed, then fossilism enters the cosmic equation with the spectre of death and extinction of the individual or of the species/civilisation.

A romantic like Blake tended to see "the deep past" of (natural/cosmic) history and "the deep present" (or "dream past") of religion, art, and science (culture) as actually forming just one holistic reality: past, present and future were all one, the paradox of *finite infinity*, a notion later asserted also by

Emily Dickinson in poem 1695. *There is a solitude of space*, and in the last analysis equivalent to Mitchell's identification of the romantic image with the composite of fossil (finite) and totem (infinity).

Among others, however, this notion, which we called "interfinitude", no doubt a romantic fundament, perhaps the most important, was asserted by many other thinkers such as: Philolaus, Nicholas of Cusa, William Blake, Gustav Fechner, Hermann Keyserling, Paul Tillich, Karl Barth, A.N. Whitehead, Charles Hartshorne, Edgar S. Brightman, James Ward, William James, Nicolai Berdyaev, George F. R. Ellis, etc., and it can be derived from the works of Schelling, Fichte and Hegel.

It would appear thus that Mitchell is right in assuming that "all roads lead to romanticism" and "from it to our own time as well" (Mitchell: 170), and in extending Bruno Latour's statement that "we have never been modern" (Latour, 1993) with his own conclusion that "this must mean we have always been romantic", because the romantic age is unparalleled in its cultural ebullience, verily a Big Bang that was in the making since human energies began to be channelled into the vision quest for a quaternary ideal: individual, social, cosmic-temporal, universal-eternal freedom, deciphered as the freedom of individual thought, the freedom of social exploration, the freedom of cosmic-temporal creation, the freedom of universal-eternal life. All these dimensions strive to compositely combine two incomensurable dimensions, the finite natural and the infinite supernatural, that is why we consider that the notion of natural-supernatural interfinitude is the most important inheritance we have from the romantics, now further decoded by Mitchell as meaning the complex fossil-totem dialectics.

It is thus significant to observe that as recently as 2002 the hypothesis of finite infinity was applied by George Ellis of Cape Town University to modern cosmology, and was declared by him to be one of the most important open key questions of modern cosmology.

The romantics' concern with the cosmic started thus from the interest in the freedoms of the individual as a being located in a deterministic universe which tends to crush his freedom. But this deterministic universe had a secret: it was not simply deterministic, it contained in itself, hidden and deeply embedded, the dimension of the infinite, the infinite room of freedom. But this room was invisible, supernatural in nature, and superposed on the visible structure. German *Naturphilosophie* developed precisely as an attempt to describe the paradoxical visible and invisible fabric of things. As Réné Wellek observed, Fr. Schelling was the father of German *Naturphilosophie*, in the last analysis a strongly metaphysical conception of nature. In Schelling's view, Nature was to be interpreted only by analogy with the progress of man towards self-consciousness. This notion the Romantics inherited from Plato's *Timaeus*, from Paracelsus and Jakob Böhme (who in turn inherited large parts of their knowledge from medieval figures such as Roger Bacon, Albertus Magnus, Raimundus Lullus, etc.), and it was established even more by Swedenborg and Blake himself. Wordsworth saw in the world of nature almost a dreamlike unreality, while Leopardi spoke of the great poetic harmony, the *effetto poetico*, of the whole of nature, almost like a process of holistic resonance of energies of all kinds. Thus, the romantics' *Naturphilosophie* became their view of an organic nature, deriving from Neoplatonism, via Giordano Bruno, Jakob Böhme, the Cambridge Platonists (such as Ralph Cudworth and Henry More), and Shaftesbury. The result was that many romantics like John Clare or William Wordsworth mythologized nature or naturalized mythology. For instance, Charles Nodier (1780-1844), a technical entomologist who was aware of Swedenborg's, Malpighi's and Saint-Martin's works, mythologized the world of insects, seeing in nature an alphabet which we needed to decipher.

In the world of insects and infusoria he saw a grotesque parallel to the forms of human art, while in the universe at large he saw a process of *syngenesis*: a fantastic evolution towards a human Utopia. Among other things, Nodier wrote a romantic fairy tale entitled *La Fée aux Miettes* (The Crumb Fairy), and romantic fantasies such as *Lydie ou la Résurrection* (1839), the latter having clear affinities with Novalis's *Hymnen an die Nacht*.

This dialectical process of *Naturphilosophie*, by which nature and mythology, nature and philosophy in its largest sense as integral science were brought into a dialogue, was a crisscrossing of liminal energies, relevantly described by Novalis himself: the ideal of the romantic *Naturphilosoph* was to render the romantic-classic, and the classic-romantic: in short, to simultaneously romanticize and classicize the world, that is, we would assume, to find the mystery of *finite infinity*, described by a Hermann Keyserling as being the golden section, and seen by Blake in the infinite of the human form ("the infinite & Eternal of the Human form", *Vala*, IX, 374), and maybe anticipated in such ideas as embraced by Thomas Campion (1567-1620), who in his poem *There is a Garden in Her Face* would equate nature with both the finite human form and the infinite of spirit.[6] Blake was later to adapt this metaphor into a mystical principle: man's natural body is a garden of delight, as he announced in *Milton* 26, 31-36:

> And every Generated Body in its inward form
> Is a garden of delight & a building of magnificence,
> Built by the Sons of Los in Bowlahoola & Allamanda:
> And the herbs & flowers & furniture & beds & chambers
> Continually woven in the Looms of Enitharmon's Daughters,
> In bright Cathedron's golden Dome with care & love & tears.

The quaternary ideal of the romantic freedoms was, however, inherited from the earlier ages of accumulation: the Romantic age was the critical age of cultural explosion coming down as a consequence of the Enlightenment, the Renaissance, the Middle Ages, etc. The Romantic man is thus a cognitive elemental Titan resting on the shoulders of previous giants, just as the man of the Enlightenment rested on the shoulders of the man of the Renaissance, who rose in turn on the shoulders of the previous man of the Middle Ages, etc.

In each age, as for instance in the Renaissance, the previous burden is partially forgotten, but not dissolved: the weight will always tend to resurface into memory in the next age. In the Enlightenment, reason rises and becomes the equal of Nature, whom it almost tends to overthrow. Now the marriage between reason and nature is a cold compact, not driven by depth of life and feeling, but downright by "calculus": Newton's fluxions and Leibniz's calculus become a mathematical tool (which an Albert of Saxony could—if he had had the chance—have dreamed about) for describing Nature into a mathematical straitjacket and while so doing Nature is divided into infinitesimal slices of reality. As Bronowski has shown, now Nature is no longer seen as something static, it becomes a process, but this is a mechanical process. Mechanism now tends to enter every description of reality, and this tendency is evidenced in science even up to the nineteenth century, after Maxwell's revolution, namely, his unification of electricity and magnetism.

The cold marriage of the Enlightenment between nature and reason forms what we could call a "Urizenic age", an age of rational-empirical man. The straitjacket of reason, however, soon stifled the huge energies of the rising intellectual powers of man, to the extent that a French philosopher like Julien Offroy de la Mettrie, in his *L'homme machine—The Machine Man* (1748), came to describe man as being a machine devoid of freedom and will, the body and the soul being mortal, while life and thinking were only the mechanical action of the nervous system.

Similar views were embraced by the associationist psychologists David Hartley and Joseph Priestly, according to whom the human psyche was governed by mechanical laws similar to those found in Nature. To be sure, such views did not end with the romantic age, on the contrary, they appear even as late as 1912, in Jacques Loeb's book *The Mechanistic Conception of Life*, in which he uses notions of physics and chemistry in order to explain the origin of life.

The culmination of this mechanistic view of life in the twentieth century was surely behaviourism, first formulated by John Watson in the *Psychological Review* in 1913. According to him, the human mind is like a "black box", a thesis which obviously reasserts Locke's view of the human mind as a tabula rasa, this doctrine being one of the most extreme forms of materialism, by virtue of which man is emptied of any spiritual residue.

It was anticipated by John Earle (1600-1665) in his *Microcosmography*, where he speaks of man's soul as "a white paper unscribbled with observations of the world, wherewith, at length, it becomes a blurred note-book." A similar view was advanced even as late as 1925 in Albert P. Weiss's *A Theoretical Basis of Human Behaviour*, in which he reduces psychology to the physical sciences. What followed after the Enlightenment with its extreme empiricism, rationalism, mechanism and materialism was the revolution of the infinite energies of love, the affections, the emotions, in the romantic age. The romantic man of feeling was born as a changed semi-replica to the platonic knight of the Middle Ages, who in turn had been born by virtue of man's instinct to protect the matrix, the spiritual principle of human regeneration. The spiritual principle of carnal perpetual postponement, however, characteristic of the medieval knights and born from the ethereal wish to render infinite the energies of love, had been overthrown into the opposite during the Renaissance and the Baroque. During the Enlightenment the principle of postponement starts its upwards surge through the obstacles of natural materiality with its principle of natural necessity, so much so that the next revolution that follows will be directed against cold reason and its tyrannous domination over the emotional life of man, but also, partially, against the tyrannous domination of necessity and so determinism, especially the mechanistic. Freedom, emotion, imagination and unity become now the four fundamentals of the romantic revolution.

The cry "Back to nature" becomes one of the generative forces of romanticism, and it derives, via Rousseau and his *Discourse on the Origin of Inequality Among Mankind* (1755), in which he first used the terms "noble savage" and "return to nature", lastly from the spirit of Native America, palpable for instance in such works as *Voltaire's L'Ingénue* (1767)—translated as *The Huron or The Pupil/Child of Nature or as The Innocent*. As Judith Nies pointed out, Enlightenment ideas about justice, equality and freedom were used to attack monarchy and aristocracy and were based on the new notion of man in nature often derived from the descriptions registered by French Jesuit missionaries who attempted to evangelize the Native Indians of Canada like the Huron (Nies, 1996). However, as Harold Bloom had pointed out, Romanticism's final goal was not unification with nature, because nature still contained the principle of necessity, even if that was not mechanistic but organic necessity. Romanticism's final goal was transcendence beyond Nature, the embrace of Spirit, because Spirit was the only realm of indeterminacy/unconditionality (as Kant had announced), while Nature still belonged to the realm of determinacy/conditionality/necessity.

The Romantic age was thus necessarily a "Centauric" age, in Ken Wilber's terminology, a marriage between mind and body, spirit and matter, an age which constitutes a temporal portal towards the transcendent, a bridge between nature and spirit. Descartes and Kant had postulated the fact that body and mind, and, respectively, phenomenon and noumenon are for ever separated, even if Descartes added the divine agent in his scheme, by which mind was able to operate on the matter from which it was separated. This deeply fragmented worldview was the Urizenic dimension of Enlightenment. The Romantics literally exploded the "gates" of Blake's Urizen or reason, thus opening the way for man's journey within, back to heaven, through the natural organic. This was a dramatic process, a titanic liberation of the huge energies of emotion, literally equatable with a human Big Bang: the huge energies of human imagination and feeling were now freed beyond measure, a deep onrush of creativity went loose into the world in an explosion of expression hardly ever to be met elsewhere in history. The romantic Big Bang comprized several dimensions:

i) freedom is its ur-phenomenon;

ii) infinity (as in the infinite abundance principle) is the favourite medium of total expression;

iii) the total work of art, and the poet as "unacknowledged legislator of the world" tend to embrace cognitively and emotionally (by virtue of a sort of Heraclitean fire of emotion) the whole of reality;

iv) holism is its password (the unity of being). The much disputed failure of Romanticism is from this viewpoint no failure at all: Romanticism as a cultural Big Bang is an undeniably explosive period when the substance of the new cultural universe is created. As such, as Margaret Atwood observed, in the post-postmodern world, i.e. in the neo-realistic, "we are still in the shadow of Romanticism." In the terms we propose, this reality is read as follows: Romanticism is the initial creative explosion, and so we are still inside the explosion, inside its wave, so that more time is needed for us to actually see clearly settling the created substance that was generated by the initial blast. If Nietzsche had said that geniuses are dangerous, we propose a different view: while it is true that such huge phenomena as romanticism are to a certain degree dangerous intellectually and otherwise, the alternative would be total lack of creativity or the illusion of creation as in the spectral worlds of a Urizen or the other Zoas in Blake's mythology. Any initial moment of creativity contains at least an utmost dangerous critical peak, but without such, creation as we know it would not exist at all. The further we get from the initial explosion of the Romantic revolution, the deeper we see into its consequences, a few of which we shall trace down and analyze in their deep logic.

Romanticism was born as a *Zeitgeist* favouring the exploration of infinity inside the finite. An impetus in that direction had been provided by the discovery of the microscope which unveiled a hidden unexpected reality. This led thinkers to surmise that there is infinite detail to life and the cosmos. Romanticism thus meant primarily the exploding of atomism: matter now was seen as being infinitely deep, unfathomable. A second impetus in the same direction came from the discovery of the telescope: the stars were very very far away; there was thus doubt as to the real dimensions of the cosmos. The question was: is the universe an infinite expanse of space or a "finite infinity", as Emily Dickinson put it in poem 1695? The question remained unanswered to this day. But what is certain is that Romanticism primarily meant the exploding of Newtonianism, with its view of the universe as a clockwork mechanism finite in nature. The cosmos became theoretically an interfinite organism, one in which the dimensions of the finite and the infinite coexist, as in Philolaos's old cosmological equation: the universe, in which Nature is embedded, in Romanticism becomes the convergence between matter and spirit. This is a fundamental romantic topos.

Owing to an inherited doctrine of correspondences between the macro- and the microcosm (derived probably from the Babylonians), the hypothesis of "finite infinity"—which we call *interfinitude*—was applied also to the human proportions/ dimensions, as well as to the nature of God. Thus, the human being became also an unfathomable, though finite, reality, and this new perception of man prepared the way to the Jungian analytical (or depth) psychology of the twentieth century, later to be projected into the interests of linguists such as Noam Chomsky regarding the deep structures of language.

A Romantic like Blake would see in man the expression of "the infinite of the human form." In other words, the microscope and the telescope exploded the "classical" notions about the dimensions of reality (atomism and Newtonianism). Reality suddenly became unfathomable: but this happened not because before reality was not so, on the contrary, it is only now that man's perceptive cognitive lens begins to observe this marvelous phenomenon of reality's unfathomableness—both in the physical sphere and in the mental-cognitive-intellectual-emotional imaginative, which presupposes a capacity to highly discern the surrounding reality, of which the closest to grasp is Nature.

If these hypotheses deriving from romantic theory be correct, then a further is derived: time itself becomes unfathomable, so much so that Romanticism prepares the way for another explosion, namely that of creationism: the earth is no longer just 6,000 years old, but much much older. As James Hutton, the founder of geology, had put it in his presentation to the Royal Society on March 7, 1785, the earth's age is indefinite. In his revolutionary *Theory of the Earth*, Hutton had argued that the Earth is incalculably old because the principle of *uniformitarianism* (which he formulated) governs the evolution of natural phenomena. According to this principle, terrestrial geology is based on the rock cycle: mountains collapse by erosion and get back to the sea, where they constitute the rocy mass of the future mountains, which again will collapse, always following the same system of arborescent mountain-ocean decomposition. One of the amazing consequences of this view was that what is now dry land was long ago the bottom of the ocean. In itself, this was not a new idea, Leibniz (1646-1716) having published already as early as January 1693 an abstract of his *Protogaea*[7] in *Acta Eruditorum* (edited in Leipzig), in which he stated that "the sea once covered a large part of the Earth which now is uncovered, and remained on the high mountains for a long time," as is evidenced by the fact that "between mountains we encounter sloping strata of remains and broken pieces, often full of shells, of glossopetrae[8] and other marine remains, captive in the hardened sediments."

The romantics must have been fascinated by the prospect of their standing on a land that in the past was the estate of oceans. Blake was professedly embracing an oceanic-aquatic consciousness by presenting space in aquatic terms. But Hutton also proposed the theory of vulcanism, according to which the origin of many rocks was volcanic. Hutton thus opposed Abraham G. Werner's theory of neptunism, which was convenient for the creationists and according to which the origin of the visible world was entirely aquatic, dry land being precipitated from the solution of a sea which had covered the entire terrestrial globe. Had neptunism been true, Hutton argued, the earth would have long become a "global swamp", because the surface of the earth is continuouosly eroding. Hutton thus posited that the structure of nature is the result of changes having taken place in very long, indefinite preiods of time. Natural events in the past can thus, according to Hutton, be explained as similar processes occurring in the present, in repetitive patterns.

Thus Hutton came to state that "we find no vestige of a beginning, no prospect of an end" in Nature (McPhee, 2000). In other words, Hutton suggested the vision of a "deep time", an unfathomable, indefinite time both in the past and the future. Hutton seems as a consequence to have played an important role in the romantic revolution, possibly by his influence through the *Oyster Club*, which he founded together with other scientists such as Joseph Black (who discovered carbon dioxide), and in which he met his friends, among whom we may mention Adam Smith, David Hume, James Watt, Benjamin Franklin, Adam Ferguson (one of the founders of the Royal Society in Edinburgh) and Robert Adam, an important architect who was a friend of the famous G.B. Piranesi (the latter dedicated an engraving and a book to Robert Adam).

Although *uniformitarianism* was attacked by the creationists, it was later popularized by Charles Lyell (1797 1875) and became a foundation of modern geology. In his remarkable book entitled *Annals of the Former World* (1998), John McPhee argued in fact that Hutton's vision is, by its effects, comparable with Copernicus's vision, namely, Hutton snatched us gradually from our privileged position in time, just as Copernicus gradually snatched us from our privileged position in space. In other words, in the wake of Hutton the Earth was no longer considered to be 6,000 years old, creationism thus making room for evolutionism, just as in the wake of Copernicus the Earth was no longer seen as the center of the universe, although the creationists had tried to maintain the validity of this idea, despite the more and more overwhelming scientific evidence. Moreover, Hutton observed that under the continually decomposing world there are the ruins of a world already having decomposed, under which yet again there are the ruins of another decomposed world. In the last analysis, Hutton's new revolutionary romantic scientific system described the continual creation-distruction equation about which romantics such as Blake, alchemy and many of the world's religions often speak. Similarly, the vision of stratified ruins reminds us of Piranesi's system. No doubt, it is well known that Piranesi had become an archetypal model for the Romantics, esp. what is now known as "black romanticism" (the Romantic tradition of Thomas De Quincey, E.A. Poe, and of authors of neo-romantic persuasion such as Alfred Kubin, Gustav Meyrink, etc.): by his works—his famous *vedute* or ruins—Piranesi reminded that the ruins we now see show us not only the past, but also the future. This phenomenon was later to be entirely assimilated by H.D. Thoreau in his strange view on time.

A previous attack on creationism, however, had come just a few years earlier, before Hutton's revolution: in 1779 Count Buffon had proposed in his *Epoque de la nature* (which represents one of the 44 volumes of the monumental *Histoire naturelle*, 1749-1788) one of the first modern anti creationist hypotheses, according to which the time that had elapsed from the beginning of Creation until his day was not only 6,000 years (as the biblical text suggested), but 75,000 years. The Earth according to Buffon came out of the Sun hot 75,000 years ago, as a consequence of the Sun's collision with a comet. Moreover, Buffon observed that certain animals retained in their anatomic structure vestigial organs, so that it was to be presumed that these had not been created by spontaneous generation, but rather are the consequence of an evolution, or rather an involution or evolution by degeneration: all organic species derived from a few primordial perfect types which evolved catamorphically towards imperfect living forms. In this sense, in 1767 Lazzaro Spallanzani had already rejected the theory of spontaneous generation—as Francesco Redi had done already in 1668. The theory of spontaneous generation had been proposed by John Needham in 1748, and earlier, in 1648, by Jan Baptista von Helmont. As can be seen, the Pendulum of History swings periodically back and forth, its speed being measured in decades or centuries: one full swing of the Pendulum, from affirmation of a theory to its negation, can take tens of years, or even hundreds, but the periodicity of this phenomenon seems to be irresistible.

Not long after the presentation of Hutton's revolutionary system a further event was to take place that would secure the rise of the new eco-morphic consciousness: in 1789, the year of the French Revolution, Gilbert White published the first officially acknowledged ecological work, entitled *Natural History* and *Antiquities of Selborne*, in which he described the fauna and flora of that area. The important rise in interest regarding the new ecological thinking is proved also by the huge number of editions of this work, namely, 200, thus it becoming one of the most popular works in the history of

England. This same year (1789) Erasmus Darwin published *The Loves of the Plants*, in which he presented, in poetic form, Linnaeus's botanic system: he thus speaks of the Ovidian metamorphoses of flowers with their floral harems, the sexual organs of the plants becoming knights and damsels. Erasmus Darwin's purpose had been to "enlist imagination under the banner of Science", and this will become one of the aims of Victorian science. The long poem *The Loves of the Plants* appeared as the part II of the volume *The Botanic Garden*, published the same year (1789) (part I here was *The Economy of Vegetation*). Both these works contain the germs of evolutionary theory. Among further predecessors to evolutionary thinking we should mention:

i) Gerolamo Cardano (1501-1576), who in his *De subtilitate rerum* (1551) and *De varietàte rerum* (1557) implies the belief in evolutionary changes;

ii) Ralph Cudworth (1617-1688), in whose works he, against seventeenth century deists, proposed the notion of "plastic life" as expressing God's method of creation, namely, "not by partial and occasional acts of creation but continuously and progressively through laws inherent in matter"[9];

iii) Gottfried Wilhelm Leibniz (1646-1716), who in his works, although he had accepted the cataclysmic theory, also embraced the idea that probably all later amphibia and land animals had descended from rudimentary marine forms (cf. *Miscellanea Berolinensia*, I, 1710, 111-112), so much so that Leibniz, as John B. Wilson observed (1965), brought the most significant contribution leading to the shift from the worldview of the eighteenth century based on the static immutable image of the Great Chain of Being—embraced among others by Aristotle, Plotinus, Milton and Pope—to the new worldview embraced by the German romantic school (Fichte, Schelling, Goethe), based on the process of endless becoming as an expression of the tendency of all beings to arrive at perfection;

iv) Immanuel Kant (1724-1804), with his volume *Allgemeine Naturgeschichte* (1755), in which he embraced Leibniz's view;

v) Jean Baptiste Robinet, in whose work entitled *De la nature*, published in 1761, living beings are said to form a continuous breachless scale of progress;

vi) Paul, Baron d'Holbach (1723-1789), who in his work entitled *Système de la Nature* (1770) he likewise unfolded an evolutionary theory;

vii) Johann Wolfgang von Goethe (1749-1832), who in his *Die Metamorphose der Pflanzen* (*Metamorphosis of Plants*) (1790) applied the doctrine of "becoming" (*das werdende*, which became a catch phrase with the German romantics) to the vegetable kingdom, and in his *Zur Naturwissenschaft überhaupt, besonders zur Morphologie* (Introduction to Natural Science and Morphology) (1817-1823) applied the same to the animal kingdom.

Robinet had contended that the Creator endowed all creatures with the internal energy to ascend on the hierarchical ladder up to man. Moreover, he presupposed that all substances contain life, even inorganic matter can evolve towards a living organism. Ilya Prigogine, the Belgian-Russian born scientist, almost 200 years later, was to reach a similar conclusion, namely, that matter itself is alive. Robinet's scale of progress has obviously strong affinities with the notion of the Great Scale/Chain of Being, according to which matter, plants and animals are hierarchically ordered, and which Voltaire attacked in his *Dictionnaire philosophique* of 1764.

The same year (1764), however, saw the publication of *Contemplation de la nature* by Charles Bonnet, in which he presented his theory of "preformation", according to which females contain in their body miniature forms of all future generations: within the woman's germ cells (eggs) there was an infinite series of preformed individuals, so much so that through the feminine the species was immortal and immutable. In other words, all possible descendents were already formed in the egg inside the woman. Bonnet's theory thus came close to Blake's later abyssal-composite theory of being. Moreover,

Bonnet responded to the challenge provided by the fossil evidence of extinct species in *La Palingénésie philosophique* (1769): here he asserted that the Earth periodically suffers universal catastrophes by which most of the living beings are destroyed, at which point the miniature forms inside the surviving women evolve in an ascendent way, moving up on the evolutionary ladder. Later, Bonnet was to state that this type of catastrophic ascensive evolution would allow inorganic matter to have life, animals to gain reason, and men to become angels.

This was another building block for the constitution of evolutionary theory, and an anticipation of such evolved ecological theories such as the Gaia hypothesis. Bonnet is thus considered to be the first to have used the term evolution in a biological context. Also, Bonnet's *Essai de psychologie* (1754) and his *Essai analytique sur les facultés de l'âme* (1760) move in the same direction as the future science of physiological psychology, practically founded (ca 1749) by David Hartley through his theory of vibrations (his study of consecutive images,[10] both visual and acoustic) and his hypothesis derived thereof that the mind and the body work always together. Moreover, Bonnet's catastrophe theory of evolution—by its idea that the living beings are directed towards increasing evolutionary complexity—partially anticipates the universal criterion of evolution as postulated by P. Glanssdorff and Ilya Prigogine in 1954, according to which a system far from equilibrium will shift towards higher complexity thus regaining a dynamic equilibrium.

However, the first prominent figure in the history of thought to have formulated the idea of evolution per se seems to have been Anaximander, who in his work entitled *On Nature* (sixth century BC) posited that life has its origins in the swampy media, later it moving to drier areas. Only 16 years before the publication of Bonnet's work, Benoit de Maillet proposed also an evolutionary theory in his *Telliamed* (1748)[11], in which he stated that the universe is full of seeds which develop into maritime beings; as the seas withdraw, some of these maritime beings evolve gradually into dry land animals. By trying to explain the nature and origin of fossils, de Maillet proposed that the Earth was not just hundreds of thousands of years old (Buffon's later hypothesis of 75,000 years falls in this category), but billions of years old, thus decisively anticipating the true value of Hutton's "deep time": de Maillet's cosmogonical vision postulated a period of time sufficiently long in order for it to present a plausible explanation in evolutionary terms. It has been thus suggested that *Telliamed* may have profundly influenced Dennis Diderot's thought and might even have inspired Charles Darwin's theory about evolution.

In 1812 Georges Cuvier contributed further to the birth of evolutionary theory by publishing his *Recherches sur ossements fossiles des quadrupedes*, in which he asserted the theory about the disappearance of certain species of animals during huge natural disasters (a theory later known as "catastrophism"). Cuvier had advanced this theory in 1799 after he had reconstituted the scheletons of gigantic reptiles (the mammoth in Paris already mentioned) which had just been discovered. He thus laid the foundations of (also comparative) paleontology, being among the first to have related the structure of fossils (animals) with their contemporary relatives, as da Vinci had done much earlier, the latter's influence in the official scientific establishment being, however, small owing to the fact that his notebooks remained unpublished for a ling time. A few years later, in 1819, William Lawrence published his *Lessons on Physiology, Zoology and the Natural History of Man*, in which he asserted that the human species is unique, but comprises more races. In the following years, Georges Cuvier popularized his catastrophe theory, according to which great natural desasters caused not only the extinction of large groups of animal species, but also the modification of the structure of the Earth.

Further on, evolutionary theory took new impetus in 1825 when Henry W. Bates, after spending 11 years in the jungles of South America, elaborated a theory about the mimetic behaviour of insects, which was later to contribute to the acceptance of the evolutionary theory embraced by Charles Darwin and Alfred Wallace. This belief was strengthened by the contribution of Max J. S. Schultze, who showed as early as

1825 that cytoplasm (then called "protoplasm") is approximately identical for all forms of life, as well as by Robert Chambers' book *Vestiges of the Natural History of Creation* (1744), in which the concept of evolution is promoted, although Chambers proposed no solution to explain this phenomenon. It was thus believed increasingly more that life must have a common origin and ground and must have evolved into the immensely diverse natural forms visible everywhere on Earth.

The idea of a *natural man*, however, had entered the influential view of the French encyclopedists Dennis Diderot and Jean d'Alembert, in their doctrine about the natural rights of man which was theoretically adopted into the 17 articles of the Declaration of the Rights of Man issued by the National Assembly of France between 20 and 26 August 1789. As Jules Michelet had pointed out, this had become the "creed of the new epoch", being profoundly influenced, among other things, by the Constitutions of some North-American states such as Virginia and New Hampshire, in their turn being inspired by the systems of natural thinking of the Native American nations, such as the Six Nations forming the great Iroquois Confederacy, which Benjamin Franklin was much interested in studying. (For instance, being fascinated with the protocols of Indian treaty-making, Franklin is known to have collected and published in 1762 thirteen such Indian treaties, made between 1736 and 1762.) Man's naturalness began thus to be "legally" established. This process came also as a reaction to the increasingly more powerful industrialization, spreading from England at a rather alert pace through the entire western world. The process of industrialization was parallel with the birth of a strongly materialistic view on reality, such as that embraced by Paul, Baron d'Holbach, vulgarized in his *Système de la Nature* (published in 1770 under the pseudonym Mirabaud), in which he attacked the foundations of Christianity and defended both materialism and atheism.

Year Zero of ecological thinking, however, can be considered the year 1866, in which Ernst Heinrich Phillip August Haeckel (1834-1919) published his *Generelle Morphologie der Organismen* (General Morphology of Organisms), in which for the first time he used the term ecology (Germ. *Ökologie* > Greek *oikos* = habitat, house; *logos* = discourse) to define "the totality of the science about the relations of the organism with the environment, which comprizes, in the largest sense, all the conditions of existence," and which two years later, in 1868, came to be associated with Carolus Linnaeus's (1707-1778) concept of the "economy of nature." Here Haeckel stated that ontogeny is a recapitulation of phylogeny, that is that the embryo in its development passes through all evolutionary stages that had occurred from the beginning of life and up to the present time. In the last analysis, this was to say that the cell contains an ancestral memory of all of life that has ever existed. This is no doubt an instantiation of holistic thinking, uncannily anticipating such contemporary theories as that embraced by Rupert Sheldrake in his *A New Science of Life—The Hypothesis of Formative Causation*, in which he describes the notion of morphogenetic field and morphic resonance. A theory similar to Haeckel's was published by Granville Stanley Hall in 1904: *Adolescence: its psychology and its relations with physiology, anthropology, sociology, sex, crime, religion and education*, in which he suggested that until maturity, a child recapitulates the whole history of the life of the race to which it belongs. This was the dawn of an abyssal systemic thinking related to living beings.

A systemic view of this kind was embraced also by the biologist George Coghill some time before the emergence of systems theory. As Fritjof Capra has shown in his *The Turning Point* (Capra, 1983: 316), Coghill contended that living beings are organized according to three patterns closely interrelated: (a) structure, i.e. organization in space; (b) function, or organization in time; and (c) mentation, i.e. the kind of organization which on the lower levels is closely related with structure and function, but which on the higher levels transcends space and time.

Mentation, in other words, is the dynamics of self-organization, being a *metafunction*, or the function of functions. In this sense, Gregory Bateson had stated in his *Steps to an Ecology of Mind* (1972) that the human mind is immanent not only inside the body, but also inside the paths and messages from outside the body (Capra, 1983: 317). In other words, the mind is to be found inside the body, but also, simultaneously, beyond the body. In systems theoretical terminology, man's total mind is to be found also in ample manifestations in which the individual mind is just a subsystem. A similar view had been long ago embraced by the western Sioux Indians with their idea that "a man's mind (*tawacin*) and soul (*wicanagi*) is all around the man and reaches a great way off" (cf. Aaron McGaffey Beede, 1912-1920, qtd. in Elaine A. Jahner, 1988: 56-57).

The idea that man has the power to expand his mind became a romantic paradigm, in which poets like Blake, Wordsworth or Shelley excelled. Likewise, the western Sioux Indians asserted that "when a man is intently using his mind for something, he calls to the use nearly all of his mind but still it is extending far away in every direction. "When a man is not using his mind, nearly all of it, leaves him and goes away on a far journey so that the man is "nearly-the-same-as-dead (*t'anuns'e*) but the mind has not left him." (cf. Beede qtd. in Jahner, id.). This is paradoxically a very keen description of the romantic voyage within, at least in the Blakean and Wordsworthian version. Also, according to the western Sioux Indians "the mind (*tawacin*) as the child grows into a man learns to strech himself far and wide. Thus the mind and soul (*tawacin wicanagi kici*) may stretch himself far away while the man is asleep or in a vision (*wihanble*) and behold many things; and may even travel to the home of the Great Spirit. (cf. Beede qtd. in Jahner, id.). This Sioux Indian doctrine uncannily resembles Wordsworth's famous description of the romantic vision quest in *Tintern Abbey*:

> that blessed mood
> In which the burthen of the mystery,
> In which the heavy and the weary weight
> Of all this unintelligible world,
> Is lightened:—that serene and blessed mood,
> In which the affections gently lead us on,—
> Until, the breath of this corporeal frame
> And even the motion of our human blood

> Almost suspended, we are laid asleep
> In body, and become a living soul:
> While with an eye made quiet by the power
> Of harmony, and the deep power of joy,
> We see into the life of things.

The conclusion of the biosystemic approach is thus that the environment is not only living, but also endowed with a mind of its own, like human beings. The stratified order of nature comprizes in this sense a holonic chain structure: the human individual mind is embedded inside the larger mind of the social system, this in turn is embedded in the mind of the ecological system, all being embedded inside the mental planetary system, itself embedded in the mental cosmic-universal system (cf. Capra, 1983: 317).

Year Zero of deep ecological thinking, on the other hand, can be considered to be the year 1973—almost 100 years after Haeckel's introduction of the *concept of ecology*—when Arne Naess began to discuss the philosophy which he called *deep ecology* or *ecosophy* in his article entitled "The Shallow and the Deep, Long-Range Ecology Movements: A Summary" (Oslo, 1973), in which he tried to describe the spiritual, more profound approach to Nature expressed in the works of Aldo Leopold and Rachel Carson, who succeeded in developing a deep ecomorphic consciousness by a more sensitive and opener attitude both towards nonhuman life and human beings, including oneself. In fact, in the important volume entitled *Deep Ecology* (1985), in which we are offered an inspired introduction to the wide question of ecosophy/deep ecology,[12] the authors Bill Devall and George Sessions consider Rachel Carson's book *Silent Spring*, published in 1962, as marking off the beginnings of "The Age of Ecology." In this sense, it is significant that this new age of ecology begins under the sign of Aldo Leopold's powerful idea about his convertion to "thinking like a mountain" (see his *A Sand County Almanac*, 1949), which points to a total man-nature identification of the kind the romantics had anticipated in their idea of subject-object unity, and which in the last analysis can be equated with the notion of Romantic empathy, expressed in such attitudes as Coleridge's, who came for instance to wonder how it would feel like to be inside the mind of a bird, an exercise of empathy which he had found difficult in the extreme (cf. Frederick Garber, 1977). Similar identifications/ empathy with natural elements can be found in the case of John Muir's *Yosemite and other parts of the Sierra*, Thoreau's *Walden Pond*, Jeffers's *Big Sur coast* or Ed Abbey's *desert country of Utah and Arizona* (Devall & Sessions: 241). On the other hand, the key to deep ecology or ecosophy in Arne Naess's initial version, which involved "a shift from science to wisdom", was the process of asking increasingly deeper questions about human life and Nature, much as was done in Socrates's maieutic method, and similar to Fr. Schlegel's idea that the quintessentially romantic means "the tendency towards an infinitely deep sense / meaning." The romantic Big Bang was precisely man's explosively energetic awakening to the reality of man's powers of imagination and fancy flowing immediately from his increased awareness of his unity with the entire cosmos through spirit.

In his book entitled *Where the Wasteland Ends* (1972), Theodore Roszack spoke of the birth of a deep ecological consciousness precisely in the following holistic terms which posit the *synergetic* (the resonance energy of life) and *syngenetic* (the common spring of all life: spirit) nature of deep ecological interrelatedness: "an awakening of wholes greater than the sum of their parts. In spirit, the discipline is contemplative and therapeutic." Moreover, deep ecology in the sense developed by Devall and Sessions embraces a philosophy that is both rational and spiritual, much as was the case of Romantic imagination as synthetic reason. Also, according to Devall and Sessions, deep ecology is a philosophy that "draws from the Earth wisdom of Native Americans and other primal cultures and that makes these approaches to wisdom relevant to contemporary, technocratic-industrial societies." Arne Naess's deep ecological approach is thus relevant because it identifies two ultimate norms, namely *self-realization* and *biocentric equality*, felt as being necessary measures in the contemporary context of the increasing ecological crisis generated by techno-mania and techno-pathology, by which life becomes increasingly more artificial, aseptic, sterile. And once again, Naess's *deep ecological self-realization* is a romantic category, developed by

C.G. Jung into a depth psychological essential process at the beginning of the twentieth century: individuation. In deep ecology or ecosophy, Naess thus posited a simple principle as foundational, namely "the greater the diversity, the greater the self-realization" (cf. Devall & Sessions: 76), much as Teilhard de Chardin had postulated before: "the more complexity, the more consciousness" (cf. Wilber, 1995: 119).

According to Fritjof Capra, the notion of rhythm, connected with the concept of resonance, is bound to play a fundamental role in the future elaborations of the new holistic worldview, given the fact that the systemic approach defines living beings as intrinsically dynamic, but also—as in the worldview of the western Sioux—as having visible forms which are nothing but stable manifestations of interior processes. Process and stability, in literature to be identified with, respectively, Romanticism and Classicism, are, however, compatible only if the constititive processes form rhythmical patterns, fluctuations, oscillations, vibrations, waves: the new systemic biology shows that fluctuations are crucially important in the dynamics of self-organization, ordered structures being born out of rhythmic patterns. In this context, F. Capra (1983: 327), analyzing the ideas in George Leonard's book *The Silent Pulse* (1981), suggested that this shift from structure to rhythm may contribute to finding a unified description of nature, because rhythmic patterns seem to be present at all levels, from the atomic to the cosmic, identities being in essence identities of rhythm, manifestations of an internal pulse, the latter representing the essence of personal identity.

In this sense, the importance of frequencies in perception has been underlined by neuropsychologist Karl H. Pribram[13], who developed a well known holographic model of the brain showing that visual perception is carried through the ocular system by an analysis of frequence patterns, while visual memory is organized like a hologram, a reason for which visual memory cannot be exactly located in the brain: like in a hologram, the whole is codified in each part. As F. Capra suggested, the holographic model has strong affinities with Geoffrey Chew's idea that subatomic particles are dynamically composed of each other in such a way that each implies all the others, and with David Bohm's idea that the implicate order shows that the whole of reality is embedded/enfolded in each of its parts, a property known as *holonomy*, no doubt reminding us of Schelling's and Novalis's idea according to which temporal unfolding is embedded in an abyssal-germinal form inside the moment. (For details on this question, see Mihai A. Stroe, *Romantismul german englez*, 2004). Modern physics thus seems to demonstrate that holonomy is a universal property, a kind of archetype of connection that had been known before only in mystical systems of thought,such as many versions of romanticism, asserting the essential unity of the universe.

Also, it is crucial at this point to emphasize the fact that the golden section has been shown to be the manifestation of an internal pulse of life. The golden section, which we identified as a foundation of romanticism, is the expression of the rhythmic pulsation of life, which is defiend as being a self-referential rhythmicity and homothetic resonance, a metaevolutional principle, i.e. a principle of the evolution of evolutions.

In this sense, Erich Jantsch, like Ilya Prigogine, had posited that the history of the universe is a succession of symmetry breaks, each symmetry break unfolding a new space-time *continuum* as a medium for the self-organization of structures. Each new symmetry break thus implies the passage to a new level of the very evolutive processiveness, so that evolution passes into metaevolution, symmetry breaks being temporal and spatial in nature. This phenomenon indicates the fact that the universe itself is alive and evolving, as in James E. Lovelock's *Gaia hypothesis* (expressed in his *Gaia: A New Look at Life on Earth*, 1979), according to which the Earth is a living autopoietic organism, which self-regulates, being endowed with a sort of hyper-conscience, similar to the Logos invoked by Romantics like Fr. Hölderlin. According to Lovelock's viewpoint, man is an inevitable part of and participant in the natural scene, who will be able to pass safely through the turbulent years of increasing ecological crisis of modern civilization, provided he should more quickly develop powers to control information and more slowly develop powers to generate energy. In Lovelock's own words:

provided man should be able to "always control the genie we have let out of the bottle", namely, the production of energy. To this crisis, explained for instance by Paul and Anne Ehrlich in their book *Extinction: The Causes and Consequences of the Disappearance of Species* (1981), deep ecology answered with a solution based on Gary Snyder's idea of "real work", that is—in Devall and Sessions's interpretation:

i) "working on ourselves";
ii) "the work of really looking at ourselves, of becoming more real";
iii) the work of gaining deep "insight that everything is connected";
iv) "a process of ever deeper questioning of ourselves";
v) the "intuition of organic wholeness" and of the "web of life" and total "interrelatedness" (the latter being an essential element emphasized by biologist Neil Evernden in his essay entitled *Beyond Ecology*); and
vi) (as in pure deep ecological thought) the fusion of the spiritual and the material aspects of reality.

This is in the last analysis an echo of the romantic Big Bang, which essentially marked a profound shift from a mechanical static worldview to a dynamic processual reticular organic view of man and the cosmos, based on the categories of:

i) infinite becoming;
ii) deep correspondence between all cosmic levels and dimensions;
iii) infinite delving into our own essence and into the quintessence of reality as a whole;
iv) organic hyperunity; and
v) materio-spiritual fusion. Many of these tenets are part of the Amerindian religious system—especially as expressed in the vision quest of the Sioux which describes the continuous making of the medicine man or *pejuta wicasa* (in the Siouan Lakota *dialect pejuta* = medicine; *wicasa* = man), as well as at least part of the Taoist, Buddhist and Christian vorldviews.

Paul Shepard, in the volume of essays *The Subversive Science: Essays Toward an Ecology of Man* (1969), edited with Daniel McKinley, indicated more fields in which a deep sense of the connection with Nature is present, namely: pre-classical Greece, Navajo religion, Romantic poetry of the eighteenth and nineteenth centuries, Chinese landscape painting of the eleventh century, Whitehead's philosophy, Zen Buddhism, the cult of the Cretan Great Mother, the ceremonies of Bushman hunters, and the medieval Christian metaphysics of light.

Ecosophy or deep ecology thus came to be considered by Devall and Sessions as being derived from the following sources:

i) the perennial philosophy, i.e. Pythagoras, Plato, and Spinoza, the latter having divided Nature into the infinite *natura naturans* (Nature as Maker-God/Substance) and the finite *natura naturata* (nature as the created beings);

ii) the pastoral/naturalist literary tradition: Rousseau, Goethe, Blake, Wordsworth, Coleridge, Shelley, Whitman, Thoreau, Emerson, D.H. Lawrence, Aldous Huxley, W. Faulkner, Wendell Berry, etc.;

iii) ecology;

iv) the new physics;

v) Christianity;

vi) feminism;

vii) the philosophy of native peoples (esp. the Native Americans);

viii) Eastern spiritual tradition;

ix) authors like Martin Heidegger, Gary Snyder, Robinson Jeffers, John Muir, David Brower.

Snyder's "real work" with all its consequences is considered by deep ecologists to be equal to "cultivating ecological consciousness", that is cultivating a "biocentric perspective" emphasizing the "inherent worth of other species besides humans", a perspective anticipated by Henry David Thoreau in his hortatory suggestion "Let your life be a friction against the machine," and, much earlier, in the thirteenth century, anticipated by St. Francis of Assissi in his notion of "a democracy of all God's creatures", based on two notions: (a) man's limitless rule of creation; and (b) man's humility. In this sense, in his *The Historical Roots of our Ecological Crisis*, Lynn White, Jr., called St. Francis "the greatest spiritual revolutionary since Christ", proposing that he become the patron saint of ecology (Devall & Sessions: 90), much as Albertus Magnus, given his immense scholarly contribution, was declared—by papal decree in 1941—the patron saint of all those who cultivate the natural sciences.

It is thus well known that St. Francis of Assissi thought that nature in its entirety was the mirror of God, leading like a ladder towards God. For him all creatures were his "brothers" and "sisters." In his famous canticle "Brother Sun, Sister Moon" he acknoledged the equality of all God's creation: Brother Wind, Brother Fire, Sister Water, Mother Earth. Even death for him became "Sister Death", while his serious illnesses had been called by him also his sisters. St. Francis of Assissi had a profound sense of brotherhood under God akin to that now professed by deep ecology, and similar to that professed by Giordano Bruno, in the 16th century, in his view of the infinite universe of vast interrelationships—in both the material and spiritual spheres—whose infinitely numerous worlds were moving on their own paths constantly propelled by their two-fold nature, namely the part-whole relationship, similar to the relation of the individual and the other worlds (Devall & Sessions: 91).

On the other hand, according to George Santayana, there is only one American writer, namely Walt Whitman, who entirely escaped anthropocentrism by expanding the democratic principle "to the animals, to inanimate nature, to the cosmos as a whole" (George Santayana, *The Genteel Tradition in American Philosophy*, speech at the University of California at Berkeley, 1911, cf. Devall & Sessions: 47). The programme of deep ecology is thus no other than the "reenchantment" of the world, decoded as the regaining of "spontaneous understanding", in the words of Isaac Watts: "In the life of spontaneity, human consciousness shifts from the attitude of strained, willful attention to koan, the attitude of open attention or contemplation." Of course, reenchantment here comes in contrast to Max Weber's postulation of "disenchantment" as a characteristic of (post)modern man. The counter-postulation was thus expressed by Morris Berman in his *The Reenchantment of the World* (1981). However, Berman drew attention to the fact that the interconnectedness envisioned by New Age thinkers and by the advocates of the new physics such as David Bohm contains a flaw, namely, the process metaphysics on which these systems are based is derived from cybernetics systems theory, which, in the last analysis, is a continuation of the Cartesian seventeenth century view of the cosmos as a machine (Devall & Sessions: 89).

No wonder, then, that a New Age thinker like Buckminster Fuller devised the notion that the whole Earth is a Spaceship—so, in the last analysis, a machine—in his famous book entitled *An Operating Manual for Spaceship Earth*, published in 1969. In passing be it mentioned that this cyber-cosmic image of the Earth became popular in the 1960s, the decade when the first space flights took place. In this book, Buckminster Fuller argues that there is no "world population problem", that all of the world's difficulties can be purely technologically solved by computers—by a combination of engineering, cybernetics systems and information theory. In this sense, Jeremy Rifkin, in his *Algeny: A New Word—A New World* (1983)[14], emphasized that now, as a consequence of the "computer/information revolution", the Darwinian view of natural evolution is being replaced by genetic manipulation carried out for man's limited purposes, resulting in the genetic metamorphosing of all life forms on Earth (Devall & Sessions: 139). In other words, New Age/ Aquarian Conspiracy thinkers talk about the "humanizing" of the Earth, a logical ramification thereof being the "humanizing" of the cosmos by breaking the "high frontier", a process to be soon started by the so-called terra-forming of the Moon, for instance. In this sense, James M. Beggs spoke, as early as 1984, about an American moon colony to be established by the year 2010.

On the other hand, spontaneity, contemplation and cognitive openness, so much treasured by deep ecologists, were for romantics like Blake, Wordsworth or Clare among the most valued powers of human imagination, and it is also true that spontaneity and individualism were major categories of freedom in the thought not only of the romantics, but also even of anarchists such as Bakunin and Kropotkin.

The *koan*, on the other hand, mentioned by Isaac Watts as fundamental for an understanding of deep ecological spontaneity, is a simple story or statement which at first sight might seem paradoxical or even nonsensical, but by looking at it from different points of view, its deep significance emerges. Precisely this process of seeing from different perspectives and depth levels is another fundamental romantic category. In this sense, Piranesi, as a predecessor of "black romanticism", gave up the principle of central perspective and so embraced more visual points. This principle of multiperspectivism was later enhanced by Fr. Hölderlin to the idea of an infinite perspective, while in W. Blake this became the stratified "Sublime Allegory" (Cf. Blake's "Letter to Thomas Butts" 6 July 1803). Moreover, in H.D. Thoreau, Piranesi's more visual points became the act of reflection revealing the infinite variability of nature: "In the reflection you have an infinite number of eyes to see for you and report the aspect of things each from its own point of view" (Cf. *The Journal of Henry D. Thoreau*, vol. XI, 1906: 213, qtd. in Stapleton, 1962: 166). For Thoreau this meant that in the act of observation we have to find a way to discover the infinity of relationships constituting actual reality. By reflection we are thus able to observe in the seemingly familiar "unsuspected lights and relations" (*Journal*, XI, 293, qtd. in Stapleton, 1962: 166). This view is indeed close to Novalis's idea of romanticizing the earthly world and logarithmizing the heavenly world, which was probably derived from the perspective of the Tabula Smaragdina, according to which what here is small is great in the heavens, whereas what here is great is small in the heavens. Also, Novalis's idea is close to the formalists' defamiliarization or *ostranenie*—making strange. In Thoreau, the justification for the need of this process was the following: "What you commonly see is but half" (*Journal*, III, 51, qtd. in Stapleton, 1962: 166), which again is a deep ecological tenet. Much as is the case of Blake's stratified "Sublime Allegory", Piranesi embedded various perspective planes on the leaf, so that the eyes of the observer be unable to find a point on which to rest, this effect corresponding to Hölderlin's Romantic *lebendige Ruhe*—living rest, and John Clare's idea of a "disorderly divine", i.e. the divine energetic rustling of Nature. In Piranesi's designs, the eyes simply glide constantly away from the already seen; beams, stairways, vaults, all get confused into each other and form a maze which generates an irresistible constant move of the eyes constituting the typical fantastic uncertainty (cf. Kleine, Bd. 36: 25-28).

This uncertainty represents a dominant in the architecture of the world depicted by such modern poets as Allen Ginsberg and by the postmodern mind in general, in the wake of Heisenberg's discovery of the principle of uncertainty or indeterminacy.

However, against the toxic results of science in the 20th century Ginsberg was to envisage a complex poetic solution, such as that presented in *Plutonian Ode*, in which we find one of the clearest (post)modern expressions of an ecological consciousness holistic and evolutionary in nature, embraced by one of America's most important disciples of William Blake.

Arguably, post-postmodernism, i.e., neo-realism, springs precisely as a consequence of the clash between postmodern fragmentariness and the continuum of globalization. The need of a neo-realism had already been anticipated by such views, close to deep ecology, as those embraced by secular theorist John Passmore in his *Man's Responsibility for Nature*, as well as by Stewart L. Udall in his *The Quiet Crisis* (1963). Thus, Passmore speaks of the need for "the emergence of a more realistic philosophy of nature" as being "the only adequate foundation for effective environmental concern", while Stewart L. Udall contends that in the twentieth century, in the decades before 1960, man began to slowly return to "some of the truths that the Indian knew from the beginning: [...] that men need to learn from nature, to keep an ear to the earth, and to replenish their spirits in frequent contacts with animals and wild land. And most important of all, we are recovering a sense of reverence for the land." Udall's prophets of this teaching were obviously Thoreau and John Muir (cf. Devall & Sessions: 59).

Also, deep ecology—as at least a branch of today's neo-realism—emphasizes the postulate of "ecological egalitarianism in principle", which seems to be a late echo or deep generalization of the Romantic principle of the French Revolution regarding freedom, brotherhood and equality. In the words of Warick Fox, an Australian philosopher, deep ecology is essentially about "the idea that we can make no firm ontological divide in the field of existence: That there is no bifurcation in reality between the human and the non-human realms [...] to the extent that we perceive boundaries, we fall short of deep ecological consciousness" (cf. Warick Fox, 1984: 194-200).

Deep ecological thinking thus takes shape as a balancing factor in the context of ecological crisis caused by the constant tendency of today's society to become over-technological. An influential book like *Animal Liberation* (1975) by Peter Singer emphasized precisely the fact that *homo technologicus* treats nonhumans more and more insensitively, which can lead to a disaster affecting not only nature, but equally mankind as embedded in Nature.

On the other hand, Erich Jantsch's idea regarding *time binding*, in which he saw a universal characteristic of evolution, was suggested to him by Alfred Korzybski's system (Korzybski, 1924, 1926; 1949). Korzybski had posed the question of a general semantics, which, according to Jantsch, could be considered as being the preparation for a cybernetic theory of living systems, given the fact that here man is seen as an organism whose behaviour operates holistically on the environment. Moreover, Korzybski's general semantics anticipates Ken Wilber's project of integral semiotics, to date a unique project in the world.

In Korzybski, time-binding referred only to the field of the human and was identified with human language as a coherent cognitive medium capable of storing and transmitting knowledge from one generation to another. Jantsch thus generalized Korzybski's hypothesis of time-binding to the level of the entire universe as a living self-organizing being possessing internal means to store and transmit its own self-knowledge. Also, Korzybski spoke of a space-binding for the case of animals: they, not possessing articulate language, cannot store and accumulate knowledge with the speed that humans do; in fact, space-binding was only referring to instinctual knowledge inherited in the animal realm from one generation to another. Jantsch changed the understanding of Korzybski's initial space-binding and time-binding hypothesis. Thus, an example of time-binding for Jantsch was the linear self-reproduction by cellular division, while the genetic transferance between bacteria was an example in which a space-binding is added.

On the other hand, Dolores LaChapelle, in her book entitled *Earth Wisdom* (1978), tried to reestablish the ancient lost ways of communicating with nature, to restore what she called "Earth bonding"—almost like an echo of Korzybski's time-binding— thus expressing her concern regarding the contemporary ecological crisis, which an author like Carolyn Merchant proclaimed as *The Death of Nature* (1980), or *ecocide.*

Arguably, the works of Alfred Korzybski, Erich Jantsch, Ilya Prigogine, James E. Lovelock, Ken Wilber and many of the specialists in ecosophy set in powerful motion the wheels of the new scientific paradigm, one of its versions being autopoietic theory, deep ecology and the science of complex systems.

All these new theories and sciences seem to point more and more clearly to the rise of a new paradigm, namely, that of *homo ecologicus* and the associated *ecotopia*, possibly of the kind envisaged by Ernest Callenbach in his novel *Ecotopia* (1975) and *Ecotopia Emerging* (1981), whose inhabitants, the Ecotopians, are philosophically modelled after the American Indian.

If Paul Shepard in his *The Tender Carnivore and the Sacred Game* (1973) announced that the present ecological crisis has been in the making for the past 10,000 years (Devall & Sessions: 172) and so it began with the agricultural revolution by which hunting and gathering were replaced by farming, being associated with man's strong desire to "completely humanize" the earth's surface, the new paradigm of homo ecologicus, as we could see in the present study, has been in the making for at least the past 1000 years, having registered a climactic wave in the Romantic Big Bang with its powerful unprecedented cry addressing man's need to reconnect to Nature, the Cosmos and Spirit, this move being an explosive counterstrike against the increasingly more visible sinking of man into materialism, empiricism, rationalism, mechanism and industrial artificiality.

The Romantic Big Bang was thus equivalent with the energetic bursting of the Blakean "mind forg'd manacles" resulting in man's reconnecting with the universal cosmic rhythms of genuine life, when Blake's mythical heroes understand that everything in the physical universe, "all you behold; tho' it appears Without, it is Within" (*Jerusalem*, 71, 18), a revelation that in fact John Muir, the American Transcendentalist, had experienced, which caused him to follow the ecological path: "I only went out for a walk and finally concluded to stay until sundown, for going out, I discovered, was actually going in" (cf. Devall & Sessions: 205).

In the theory we propose, the Romantic Big Bang reconstructed or reconfigured, at least partially, time-binding (language/ cognitive powers), space-binding (instinct, nativism), earth-bonding (planetary and cosmic consciousness) and spirit-binding (freedom-unconditionality), all of which had kept on being fragmented possibly since the advent of the agricultural revolution, as Paul Shepard had suggested.

That we now still live in the shadow of romanticism, as, among others, Harold Bloom and Margaret Atwood pointed out, is only relevant for the present approach by which we mean to attract attention to the reality that the emerging new paradigm of homo ecologicus is a necessity if man is to keep on dreaming about a future of his race safe from the spectre of atomic warfare, from "the shadow of nuclear holocaust" which we now are living under (Devall & Sessions: 207), a future that, why not, should draw from the infinite hidden energies of past generations of philosophers, scientists, artists, saints, shamans, etc, all building up an interrelated system of thought with a view to forming a perfect knowledge of life—of the kind today's *ecosophy* is striving to attain, of the kind Ramon Lull had hoped for while thinking to discover the "magnificent art", and Leibniz envisaged while dreaming about the *mathesis universalis*—potentially having the power to revert today's ecocide into a universal feast of life.

Notes and References

1. This artilce has been presented in a simplified form as a key note speech in the framework of the International Conference *Poetic Ecologies: Nature as Text and Text as Nature in English-Language Verse*, Université Libre de Bruxelles, May 14-17, 2008, and is part of my Fulbright advanced research project entitled *The Pendulum of History: Paradigm Shifts in Literature and Science from the Romantic to the Postmodern Age*, on which I worked at Yale University in 2007, under the scholarly advisorship of Prof. Harold Bloom and under the supervision of Prof. Paul Fry.
2. "I sat with love by pasture stream / Aye beautys self was sitting bye / Till fields did more then edens seem."
3. "Who will in fairest book of Nature know / How virtue may best lodged in beauty be, / Let him but learn of love to read in thee, / Stella."
4. He called these kinds of fossil organisms using Latin denominations: *ostracites* = fossil oysters; *strombites* = fossil gastropods; *trochites*, *encrinos*, *belemnites*, *lapis judaicus*, *lapis eislebanus*, etc.
5. Ojibway is an Algonquian language spoken by the Ojibwa people.
6. "There is a Garden in Her Face / Where roses and white lilies grow; / A heavenly paradise is that place, / Wherein these pleasant fruits do flow."
7. *Protogaea* was written by Leibniz between ca 1690-1693, and it was published only posthumously in 1749.
8. "Glossopetrae" are fossilized shark-teeth; lit. "tongue-stones."
9. See the New England Transcendentalist James Freeman Clarke who emphasized this idea in a review for *Galaxy*, published in December 1874. As a consequence thereof, Clarke saw no fundamental conflict between the theory of evolution and that of divine creation: "Thus, God may be regarded as creating each plant, while he maintains the mysterious force of development by which it grows from its egg or its seed." (Cf. John B. Wilson, 1965: 289).
10. I.e. those images that still remain present (as residual vibrations) on the retina / eardrum, after the visual / acoustic stimulus ceased.
11. This work was published posthumously in 1748 by Abbé Le Mascrier, who dedicated it to Cyrano de Bergerac and attributed it to a Hindoo philosopher whose name allegedly was "Telliamed": one easily notices that we are dealing with the inversion of the name of the author: "demailleT".

12. See also at least the following anthologies which constitute excellent introductions to this field: Carolyn Merchant, editor, *Ecology*, 2008; Cheryll Glotfelty & Harold Fromm, editors, *The Ecocriticism Reader*, 1996; George Sessions, editor, *Deep Ecology for the 21st Century*, 1995. For an initiation into the essential ecological language and concepts, see at least Ernest Callenbach, *Ecology: A Pocket Guide*, 2008. For an introduction to the comparative study of ecology and literature, see S. Murali, editor, *Nature and Human Nature: Literature, Ecology, Meaning*, 2008.
13. In "Holonomy and Structure in the Organization of Perception", 1977; and in "Holographic Memory", 1979.
14. *Algeny*: a term invented by Rifkin by which he understands "the upgrading of existing organisms and the design of wholly new ones with the intent of «perfecting» their performance." "Algeny" is thus an ascensive evolution which expresses an underlying entelechial principle: we reach external functional perfection by internal functional perfection (via natural evolution or by genetic manipulation, etc).

Works Cited

1. Arthur Amiotte, "The Lakota Sun Dance: Historical and Contemporary Perspectives", in Raymond J. DeMallie, Douglas R. Parks, editors, *Sioux Indian Religion: Tradition and Innovation*, 1988 [1987], pp. 75-89.
2. Margaret Atwood, *Negocierea cu moartea: Un scriitor despre scriiturã*, Tritonic, Bucuresti, 2007 [2002: *Negotiating with the Dead: A Writer on Writing*, O. W. Toad Ltd., The Press Syndicate of the University of Cambridge].
3. Georgius Agricola (Georg Bauer), *De Natura Fossilium*, Dover Publications, 2004 [1955: Geological Society of America; the original in Latin: 1546, Basileae].
4. Gregory Bateson, *Steps to an Ecology of Mind: Collected Essays in Anthropology, Psychiatry, Evolution, and Epistemology*, University of Chicago Press, Chicago, 2000 [1972].
5. J. Christian Bay, *Conrad Gesner (1516-65), the Father of Bibliography: An Appreciation*, Bibliographical Society of America, 1916.
6. Vincent of Beauvais, *The Mirrour of the World*, Transl. William Caxton (1481), Allen Press, Kentfield, California, 1964.
7. Aaron McGaffey Beede, *Journals and Letters*, Orin G. Libby Manuscript Collection, University of North Dakota, Grand Forks, 1912-1920.
8. Morris Berman, *The Reenchantment of the World*, Cornell University Press, Ithaca, New York, 1981.

9. William Blake, *Cãrþile profetice, Vala sau Cei patru Zoa*, (*Opere I*), Mihai A. Stroe, editor, Institutul European, Iaºi, 2006.
10. William Blake, *Çãrþile profetice iluminate, Milton*, (*Opere II*), Mihai A. Stroe, editor, Institutul European, Iaºi, 2006.
11. *Blake Complete Writings with variant readings*, Geoffrey Keynes, editor, Oxford University Press, Oxford, New York, 1979.
12. Jacob Bronowski, *The Ascent of Man*, Macdonald Futura Publishers, British Broadcasting Corporation, London, 1981 [1973: Science Horizons Inc.]
13. Ernest Callenbach, *Ecology: A Pocket Guide*, Revised and Expanded, University of California Press, Berkeley, Los Angeles, London, 2008 [1998].
14. Ernest Callenbach, *Ecotopia: The Notebooks and Reports of William Weston*, Bantam Books, New York, Toronto, London, Sydney, Auckland, 1990 [1975: Banyan Tree Books].
15. Ernest Callenbach, *Ecotopia Emerging*, Banyan Tree Books in association with Heyday Books, Berkeley, California, 1981.
16. Fritjof Capra, *The Turning Point: Science, Society, and the Rising Culture*, Flamingo (Imprint of Harper Collins Publishers), London, 1983 [1982: Simon & Schuster, New York].
17. Rachel Carson, *Silent Spring*, Mariner Books / Houghton Mifflin Company, New York, N.Y. 2002 [1962].
18. Robert Chambers, *Vestiges of the Natural History of Creation, and Other Evolutionary Writings*, University of Chicago Press, Chicago, 1994 [1844: London].
19. John Clare, *Poeme*, Mihai A. Stroe, editor, Institutul European, Iaºi, 2007.
20. George Crowder, *Anarhismul clasic: Gîndirea politicã a lui Godwin, Proudhon, Bakunin ºi Kropotkin*, Antet, Oradea, 1997 [1991: *Classical Anarchism*, Oxford University Press].
21. Erasmus Darwin, *The Botanic Garden: A Poem in Two Parts:* Part I, *The Economy of Vegetation*; Part II, *The Loves of the Plants*, J. Jackson, London, 1789, 1791.
22. Benoit de Maillet, *Telliamed, or Conversations Between an Indian Philosopher and a French Missionary on the Diminution of the Sea*, Transl. Albert V. Carozzi, University of Illinois Press, Urbana, 1968 [1748: Abbé Le Mascrier].
23. Raymond J. DeMallie, Douglas R. Parks, editors, *Sioux Indian Religion: Tradition and Innovation*, University of Oklahoma Press, Norman, London, 1988 [1987].
24. Raymond J. DeMallie, "Lakota Belief and Ritual in the Nineteenth Century", in Raymond J. DeMallie, Douglas R. Parks, editors, *Sioux Indian Religion: Tradition and Innovation*, 1988 [1987], pp. 25-43.

25. Deutsche Akademie der Wissenschaften zu Berlin, *Miscellanea Berolinensia ad Incrementum Scientiarum, Ex scriptis Societati Regiae Scientiarum exhibitis edita, cum figuris aeneis et indice materiarum*, Johan. Christ. Popenii, Berlin, 1710-1744.

26. Bill Devall, George Sessions, *Deep Ecology*, Gibbs M. Smith Inc., Peregrine Smith Books, Salt Lake City, 1985.

27. Paul H. T. d'Holbach, *Système de la nature, ou Des loix du monde physique et du monde moral*, Partie I, Elibron Classics, Adamant Media Corporation, 2005 [1780: London].

28. Emily Dickinson, *The Complete Poems of Emily Dickinson*, Thomas H. Johnson, editor, Little, Brown and Company, New York, Boston, London, 1961 [1960].

29. Emile Durkheim, *The Elementary Forms of Religious Life*, Transl. Carol Cosman, Oxford University Press, Oxford, 2008 [2001] [1912].

30. Paul & Anne Ehrlich, *Extinction: The Causes and Consequences of the Disappearance of Species*, Ballantine, New York, 1981.

31. Neil Evernden, "Beyond Ecology: Self, Place, and the Pathetic Fallacy", in Cheryll Glotfelty & Harold Fromm, editors, *The Ecocriticism Reader: Landmarks in Literary Ecology*, University of Georgia Press, Athens, Georgia, 1996, pp. 92-104.

32. Neil Evernden, *The Natural Alien: Humankind and the Environment*, University of Toronto Press, Toronto, 1985.

33. Neil Evernden, *The Social Creation of Nature*, The Johns Hopkins University Press, Baltimore, Maryland, 1992.

34. Warick Fox, "Deep Ecology: A New Philosophy of Our Time?", in *The Ecologist*, v. 14, 5-6, 1984, pp. 194-200.

35. Sigmund Freud, *Totem und Tabu*, Fischer, Frankfurt. 1991 [1913].

36. R. Buckminster Fuller, *An Operating Manual for Spaceship Earth*, Aeonian Press, Mattituck, N.Y., 1978 [1969: Southern Illinois University Press].

37. Frederick Garber, "Nature and the Romantic Mind: Egotism, Empathy, Irony", in *Comparative Literature*, Vol. 29, No. 3., Summer, 1977, University of Oregon, pp. 193-212.

38. Cheryll Glotfelty & Harold Fromm, editors, *The Ecocriticism Reader: Landmarks in Literary Ecology*, University of Georgia Press, Athens, Georgia, 1996.

39. Ernst Haeckel, *Generelle Morphologie der Organismen*, Gruyter, 1988 [1866].

40. Granville Stanley Hall, *Adolescence: Its Psychology and its Relations with Physiology, Anthropology, Sociology, Sex, Crime, Religion and Education*, Volume 1, Elibron Classics, Adamant Media Corporation, 2005, [1908: Appleton & Co., London].

41. Gerold Hayer, *Konrad von Megenberg*, Das Buch der Natur: *Untersuchungen zu seiner Text- und Überlieferungsgeschichte*,

Münchener Texte und Untersuchungen zur deutschen Literatur des Mittelalters, M. Niemeyer, 1998.

42. Friedrich Heer, *The Medieval World, Europe 1100-1350*, Phoenix Giant, London, 1998 [1961].
43. Werner Heisenberg, *Pa°i peste granipã*, Transl. Ilie Pârvu, Editura Politicã, Bucure°ti, 1977 [1971: *Schritte über Grenzen, Gesammelte Reden und Aufsätze*, C. R. Piper & Co. Verlag, München].
44. Alexander Hellemans & Bryan Bunch, *Istoria descoperirilor °tiinpifice*, Editura Orizonturi, Bucure°ti, ca. 1995 [1988: *The Timetables of Science: A Chronology of the Most Important People and Events in the History of Science*, Simon & Schuster, New York].
45. James Hutton, *Theory of the Earth (1788)*, Nu Visions Publications, LLC, Sioux Falls, SD, USA, 2007 [1788: *Theory of the Earth, or An Investigation of the Laws Observable in the Composition, Dissolution, and Restoration of Land upon the Globe*, Edinburgh].
46. Elaine A. Jahner, "Lakota Genesis: The Oral Tradition", in Raymond J. DeMallie, Douglas R. Parks, editors, *Sioux Indian Religion: Tradition and Innovation*, 1988 [1987], pp. 45-65.
47. Erich Jantsch, *The Self-Organizing Universe*, Pergamon, New York, 1980.
48. Erich Jantsch, *Design for Evolution, Self-Organization and Planning in the Life of Human Systems*, George Braziller, New York, 1975.
49. Immanuel Kant, *Universal Natural History and Theory of the Heavens*, Transl. Stanley L. Jaki, Scottish Academic Press, Edinburgh, 1981 [1755: *Allgemeine Naturgeschichte und Theorie des Himmels*].
50. Frank Kermode & John Hollander (general editors), Harold Bloom, Martin Price, J. B. Trapp, Lionel Trilling (editors), *The Oxford Anthology of English Literature*, 2 vol., Oxford University Press, London, New York, Tornoto, 1973.
51. Sabine Kleine, *Häßliche Träume, Literarische Phantastik und das anti-ästhetische Projekt der Moderne*, Schriftenreihe und Materialien der Phantastischen Bibliothek Wetzlar, Bd. 36.
52. Alfred Korzybski, *Time-Binding: The General Theory*, Two Papers, 1924-1926, The Institute of General Semantics, Lakeville, Connecticut, 1971 [1924, 1926, 1949].
53. Arthur Koestler, *The Act of Creation*, Arkana, Penguin Books, London, 1989 [1964: Macmillan, New York].
54. Dolores LaChapelle, *Earth Wisdom*, Way of the Mountain Center, Silverton, Colorado; Finn Hill Art, 1984 [1978: Guild of Tudor Press]
55. Bruno Latour, *We Have Never Been Modern*, Transl. Catherine Porter, Harvard University Press, Cambridge, Massachusetts, 2001, [1993: Harvester Wheatsheaf; 1991: La Découverte, *Nous n'avons jamais été moderne*].

56. Gottfried Wilhelm Leibniz, *Protogaea*, Transl. Claudine Cohen & Andre Wakefield, University of Chicago Press, Chicago, 2008 [1749: *Protogaea, sive, De prima facie tellvris et antiqvissimae historiae vestigiis in ipsis natvrae monvmentis dissertatio; ex schedis manvscriptis viri illvstris in lvcem edita a Christiano Ludovico Scheidio*, Svmptibus Ioh. Gvil. Schmidii, Göttingen].

57. George Leonard, *The Silent Pulse: A Search for the Perfect Rhythm that Exists in Each of Us*, Revised & Updated for the 21st Centruy, Gibbs Smith, Publisher, Layton, Utah, 2006 [1981: Bantam, New York; 1980: Wildwood House, London].

58. Aldo Leopold, *A Sand County Almanac, With Essays on Conservation*, Oxford University Press, London, 2001 [1949].

59. John Long, *Voyages and Travels of an Indian Interpreter and Trader: describing the manners and customs of the North American Indians; with an account of the posts situated on the river Saint Laurence, Lake Ontario, &c. to which is added a vocabulary of the Chippeway language, names of furs and skins, in English and French, a list of words in the Iroquois, Mohegan, Shawanee, and Esquimeaux tongues, and a table, shewing the analogy between the Algonkin and Chippeway languages*, London, 1791 (A. H. Clark, Cleveland, 1904).

60. Arval Looking Horse, "The Sacred Pipe in Modern Life", in Raymond J. DeMallie, Douglas R. Parks, editors, *Sioux Indian Religion: Tradition and Innovation*, 1988 [1987], pp. 67-73.

61. Konrad Lorenz, *Das Sogenannte Böse: Zur Naturgeschichte der Aggression*, Verlag Dr Borotha-Schoeler, Wien, 1963.

62. James E. Lovelock, *Gaia: A New Look at Life on Earth*, Oxford University Press, Oxford, 1979.

63. Albertus Magnus, *De Mineralibus*, Joannes and Gregorius de Gregoriis, Venice, Italy, 1495 [1476].

64. John McPhee, *Annals of the Former World*, Farrar, Straus and Giroux, New York, 2000 [1998].

65. Carolyn Merchant, *The Death of Nature: Women, Ecology, and the Scientific Revolution*, Harper One, An Imprint of Harper Collins Publishers, New York, 1990 [1980].

66. Carolyn Merchant, editor, *Ecology: Key Concepts in Critical Theory*, 2nd edition, HB, Humanity Books, An Imprint of Prometheus Books, Amherst, New York, 2008.

67. W. J. T. Mitchell, "Romanticism and the Life of Things: Fossils, Totems, and Images", in *Critical Inquiry*, Vol. 28, No. 1, Things, Autumn, 2001, The University of Chicago Press, pp. 167-184.

68. *Monthly Review*, June, 1792, review of John Long's book *Voyages and Travels of an Indian Interpreter and Trader*, London, 1791, pp. 129-137.

69. Arne Naess, "The Shallow and the Deep, Long-Range Ecology Movements: A Summary", in *Inquiry*, 16, Oslo, 1973, pp. 95-100.

70. John M. Nicholas, editor, *Images, Perception and Knowledge*, Reidel, Dordrecht, Holland, 1977.

71. Judith Nies, *Native American History: A Chronology of a Culture's Vast Achievements and Their Links to World Events*, Ballantine Books, New York, 1996.

72. George Ochoa & Melinda Corey, *Ghidul cronologic al ºtiinþei*, Editura All Educational, Bucuresti, 2000 [1995: *The Timeline Book of Science*, The Stonesong Press].

73. John Passmore, *Man's Responsibility for Nature: Ecological Problems and Western Traditions*, Gerald Duckworth & Company, London, 1974.

74. Sherman Paul, editor, *Thoreau, A Collection of Critical Essays*, Prentice-Hall, Inc., Englewood Cliffs, N.J., 1962.

75. Ioan Aurel Preda, *English Romantic Poetics*, Institutul European, Iasi, 1995.

76. Karl H. Pribram, "Holonomy and Structure in the Organization of Perception", in John M. Nicholas, editor, *Images, Perception and Knowledge*, Reidel, Dordrecht, Holland, 1977.

77. Karl H. Pribram, " 'Holographic Memory'. Interview by Daniel Goleman", in *Psychology Today*, February, 1979.

78. Karl H. Pribram, *Brain and Perception: Holonomy and Structure in Figural Processing*, Lawrence Erlbaum Association, New Jersey, 1991.

79. Jeremy Rifkin (with Nicanor Perlas), *Algeny: A New Word – A New World: A Provocative Critique of Darwinism*, Viking, New York, 1983.

80. Theodore Roszack, *Where the Wasteland Ends: Politics and Transcendence in Postindustrial Society*, Doubleday, Garden City, N.Y., 1972; Anchor Books / Doubleday, Garden City, N.Y., 1973.

81. Julius Ruska, *Tabula Smaragdina: Ein Beitrag zur Geschichte der Hermetischen Literatur*, C. Winter's Universitätsbuchhandlung, Heidelberg, 1926.

82. S. Murali, editor, *Nature and Human Nature: Literature, Ecology, Meaning*, Prestige Books, New Delhi, 2008.

83. George Sessions, editor, *Deep Ecology for the 21st Century: Readings on the Philosophy and Practice of the New Environmentalism*, Shambhala Publications, Inc., Boston, Massachusetts, 1995.

84. Rupert Sheldrake, *A New Science of Life – The Hypothesis of Formative Causation*, J. P. Tarcher, L.A., 1981.

85. Paul Shepard & Daniel McKinley, editors, *The Subversive Science: Essays Toward an Ecology of Man*, Hougton Mifflin Company, Boston, 1969.

86. Paul Shepard, *The Tender Carnivore and the Sacred Game*, University of Georgia Press, Athens, Georgia, 1998 [1973: Scribners, New York].

87. Peter Singer, *Animal Liberation*, Harper Collins Publishers, New York, N.Y., 2002 [1975, 1990].

88. Gary Snyder, *The Real Work: Interviews and Talks: 1964-1979*, New Directions Pub. Corp., New York, 1980.

89. Laurence Stapleton, "Introduction", in Sherman Paul, editor, *Thoreau, A Collection of Critical Essays*, Prentice-Hall, Inc., Englewood Cliffs, N.J., 1962.

90. Claude Lévy-Strauss, *Totemism*, Beacon Press, Boston, 1971 [1963; French original: 1962, Presses Universitaires de France].

91. Mihai A. Stroe, *John Clare °i canonul romantic: Metafizica Naturiisi poetica Logosului mitic*, in John Clare, *Poeme*, Mihai A. Stroe, editor, Institutul European, Ia°i, 2007, pp. 5-104.

92. Mihai A. Stroe, *Romantismul german °i englez. ªtiinþa arhetipurilor, ipoteza interfinitudinii °i numãrul de aur*, Institutul European, Ia°i, 2004.

93. Bradford Torrey & Francis H. Allen, editors, *The Journal of Henry D. Thoreau*, vol. III and XI, Houghton Mifflin Company, Boston, 1906.

94. Hermes Trismegistus, *Tabula Smaragdina*, Editura Herald, Bucuresti, 2004

95. Stewart L. Udall, *The Quiet Crisis*, Holt Rinehart & Winston, New York, 1963 (*The Quiet Crisis and the Next Generation*, Peregrine Smith Books, Salt Lake City, 1988).

96. Gilbert White, *Natural History and Antiquities of Selborne, in the County of Southampton: With Engravings, and an Appendix*, T. Bensley, B. White & Son, London, 1789.

97. Ken Wilber, *Sex, Ecology, Spirituality – The Spirit of Evolution*, Shambhala, Boston & London, 1995.

98. John B. Wilson, "Darwin and the Transcendentalists", in *Journal of the History of Ideas*, Vol. 26, No. 2, Apr. - Jun., 1965, University of Pennsylvania Press, pp. 286-290.

4

Ecocentric Dimensions in the Poetry of Robinson Jeffers

— Reena J. Andrews

Confronted with the realization that his actions cause irrevocable damage to the ecosphere, man's dilemma is that there are no easy solutions to the problem of the environment. It has become necessary to question the very foundations of the modern, industrial, growth-oriented way of life, which has engendered the crisis, and to shift to more viable alternatives that would deconstruct the accepted power-oriented categories of belief. A questioning of hierarchies of dominance and power—patriarchal, racist, colonial, anthropocentric—has resulted in a tentative yet gradual shift of significance from such power centres to hitherto marginalized voices that show a willingness to expound feelings of mutuality and interdependence.

Anthropocentrism, which sees humans as the most significant aspect of the cosmos, and grants only a relative significance to all other biotic as well as abiotic components, has contributed in no small measure to the delusion of human self-importance. The feeling that he stands apart from nature and is superior to other forms of life has generated in man a callous indifference to the non-human elements of the ecosphere, and resulted in his blind destruction and exploitation of the planet. On the other hand, ecocentrism assumes the intrinsic value of all natural systems, thereby effectively relegating humanity to just one among many components that make up a complex web of interrelationships and interdependencies.

In *Imagining the Earth*, John Elder describes poetry as the edge between mankind and nonhuman nature, providing an access for culture into a world beyond its preconceptions. It is such a mediating role that informs Robinson Jeffers's attempt at a vision of inhumanism (210). Jeffers's philosophical alternative to humanism, exhorting man to mature past the puerile efforts to subdue nature and attain a state of greater awareness of and concern for the physical world, has placed him in a salient position within the contemporary environmental movement. Living in troubled times, Jeffers offers in place of outworn values, a new ethic that would be less injurious to man and his environment. He chose the term 'inhumanism' for a compelling vision which projects a state of mind necessary in an era of environmental anxiety. While his long poems are intense and impassioned studies of the inhumanist outlook, "a shifting of emphasis and significance from man to not-man; the rejection of human solipsism and a recognition of the transhuman magnificence" (*The Double Axe*, xxi), the short poems also argue for a more meaningful relationship with the environment and reveal overtones that distinctly concur with the concepts of deep ecology. Bill Devall and George Sessions in *Deep Ecology* admit that the writings of poets like Robinson Jeffers contributed greatly to the deep ecology perspective (80).

The profound relationship between man and the landscape is developed in practically all Jeffers's poems. The hierarchy of significance from landscape to animal to human is in keeping with the poet's philosophy of inhumanism. Attempting to decentre humanity, Jeffers gives the non-human inhabitants of the planet their due importance and relegates the human elements to the sidelines. He insists that all life has value in itself, independent of its usefulness to man. In "Animals," he celebrates the non-human aspects of the universe that are able to survive in extremely hostile environments. Sea-lions survive where man cannot: "life near kin to human, intelligent, hot-blooded, idle and singing," floating at ease in the ice-cold water. Even the tongues of flame on the surface of the sun which assume animal-like characteristics—"the beautiful passionate bodies of living flame, batlike flapping and screaming,/Tortured with burning lust and acute awareness" (3: 364)—are more intensely alive than man can ever be, and are more worthy of praise than human beings. The poet stresses the kinship of the non-human elements to the human component, before abruptly relegating the human to a marginal position. This ability to present the universe as all alive and interrelated encourages ecological sensitivity by putting mankind in its place as one form of life among many, and certainly not the most significant.

Poems like "Original Sin" illustrate the viciousness of human beings, "the most repulsive of all hot-blooded animals," who roast a mammoth to death in a perverted desire to witness its agony. In "The Inquisitors" earth spirits examine some human specimens and wonder how such fragile creatures could be so noxious. The insignificance of humans is compared to the immensity of nature. In nature's grasp, they are "too helpless even to scream." Puzzled over man's destructive tendencies, the nature spirits on further examination discover that the human brain is just a "drop of marrow," and wonder, " 'How could that spoil the earth?' " They conclude that the destructive power is in the human discoveries. "'The blasts and the fires are nothing: freckles on the earth: the emanations/ Might set the whole planet into a tricky fever/And destroy much'" (3: 210). But in the process of destruction, human beings would not just destroy themselves, but would wipe out all life from the earth. The earth spirits finally end on a typically Jeffersian note of hope that whatever be the destruction wrought by the human race, "It is not likely they can destroy all life: the planet is capacious/ Life would surely grow up again" (3: 210).

Jeffers emphasizes the enormity of the human impact on the earth, which is rapidly getting worse, and points out that human lifestyles and the population explosion are key elements of this impact. In "Post Mortem," the poet prophesies the destruction of natural strength and beauty, as a result of over-population and exploitation. "Passenger Pigeons" reflects on species that have become extinct after explosions of population-the passenger pigeons and the American bison. The poet apostrophizes death: " 'You Death, you watch for these things, / These explosions of life: they are your food, / They make your feasts' " (3: 435). Now perhaps death has turned his grossly craving black eyes on humanity. The poet comments on mankind's seemingly impressive achievements:

> "we have bridled the cloud-leaper lightning,
> snatched the live thunderbolt/Out of God's hands."
>
> we have taken
> The primal powers, creation and annihilation; we make
> new elements, such as God never saw,
>
> We can explode atoms and annul the fragments, nothing
> left but pure energy, we shall use it
>
> In peace and war.

He compares man's misuse of his mind's abilities to the monstrous growth of the dinosaurs that had ruled the earth with tearing teeth and enormous bodies plated with armour that made them invincible. The sabre-toothed tigers had huge fangs that were as unnecessary as our sciences. Our minds are like those tusks—"hypertrophied and terrible." For the human race the time for extermination seems to be drawing near; as Jeffers comments in "Birth and Death," "Breeding like rabbits we hasten to meet the day" (3: 440).

"Not Our Good Luck" contrasts the folly of the trappings of a decadent urbanized civilization with "the ancienter simple and silent tribes of the stars." God as energy pervades the entire universe, be it the squalor of the present or the grandeur of the past. In spite of her gods and her religions, "God also moved on the city;" the poet stresses the human realm, which is seen as a dimension of the natural. Though man turns away from it, the beauty of things touches everyone; "he is nowhere not present, his beauty, it is burning in the midland villages/ And torture men's eyes in the alleys of cities. (1:13). "Thus, held together by the interweaving current of energy that flows through the universe interconnecting us with our environment and with other beings, nature becomes a single reality.

In "The Coast Road," the poet surveys the rottenness of cities from the isolation of his mountainous retreat. John Elder in Imagining the Earth describes this as a characteristic circuit in Jeffers's poetry: love of the wilderness and hatred of civilization feeding on each other. The polarization of his own feelings is further intensified for Jeffers by his perception of a larger historical process of estrangement.(8). Though the destruction that the poet portrays is still in the future, the human capacity for subduing nature seems to have acquired a frightening momentum. The desire to protect the world from men seems to translate itself into a vision of human obliteration. By separating humans from his poetic landscape, he goes beyond humanity to envision an ideal though seemingly impossible state where the non-human and the human are equal participants.

In *The Cliffs of Solitude*, Robert Zaller states that the failure to sustain cultural purpose and value marked the onset of decadence. The decadence of the West-or, more precisely, its incipient decline-emerges as a dominant theme in Jeffers' verse in the twenties and remains so to the end of his career. In "The Broken Balance," Jeffers writes:

> The world sickens with change, rain becomes poison,
> The earth is a pit, it is time to perish.
> The vines are fey, the very kindness of nature
> Corrupts what her cruelty before strengthened.
> When you stand on the peak of time it is time to begin to
> perish.

The poem is a realization that the corruption was inherent in man, not in the world around him. The balance has been broken between man and his sustaining beliefs as well as between man and his environment. It is when man starts losing his organic connection to the earth that he starts perceiving it as an object of conquest in his obsession with what he calls "progress." Robert Zaller describes the delusion of man's progress:

> Progress was the ultimate project of the modern will, the reduction of the world to terrain, of universal process to naive recurrences that, once deciphered, passively awaited man's disposal. But progress itself was curiously afflicted by loss of effect, increasing mastery by loss of control, as purpose degenerated into mere persistence and finally

> "deep indifference," a mechanized somnambulism in which the will, having lost even the memory of its own command, drifted destructively in a phantasmal void. Lost amid conquest, modern men were dwindled and anxious; in the mass they seemed still more vulnerable, as they clung to each other for support. (207)

In the poem "The Purse-Seine," Jeffers describes the fishermen going out at night in their boats to cast their huge seines for schools of fish, and he paints a realistic picture of their prey caught in the net beneath the boat:

> I cannot tell you
> How beautiful the scene is, and a little terrible, then, when the crowded fish
> Know they are caught, and wildly beat from one wall to the other of their closing destiny the phosphorescent
> Water to a pool of flame (...) (2: 517)

As he looks down from a mountaintop at night on a "wide city, the colored splendor, galaxies of light," the comparison of the people in the city to the sardines caught in the net seems to him both beautiful and terrible.

The "net" of the city separates men from their natural environment; deprived of their roots, men are helpless and unable to stand alone. They are unnaturally dependent on their fellow creatures; but that relationship is the unreal one between trapped victims; their glow is the shimmer of decay:

> I thought, We have geared the machines and locked all together into interdependence;
> we have built the great cities; now
>
> There is no escape. We have gathered vast populations incapable of free survival, insulated
>
> From the strong earth, each person in himself helpless, on all dependent. The circle is closed, and the net
>
> Is being hauled in. They hardly feel the cords drawing, yet they shine already. (2: 518)

The image of the net recurs in "The Tower Beyond Tragedy"—"the net of desire/Had every nerve drawn to the center, so that they writhed like a full draught of fishes, all matted / In the one mesh" (1: 176) —to describe a culture built on self love. Zaller describes urbanization as the outer symptom, the public manifestation of this collapse upon the self. "The tightening web of community created isolation within dependence, enlarging the sense of self while destroying the scope of free activity, as narcissism and anomie reinforced each other in a self-perpetuating cycle" (207).

Against such a background of decadence caused, to a large degree, by loss of contact with the earth, deep ecology calls for a return to a way of life that is totally tied to the rhythms of the earth. For Jeffers, the present time is a disturbed one with "misery and riches, civilization and squalid savagery,/Mass war and the odor of unmanly peace":

> In order to value this fretful time
> It is necessary to remember our norm, the unaltered passions,
> The same-colored wings of imagination,
> That the crowd clips, in lonely places new-grown; the unchanged
> Lives of herdsmen and mountain farms'
> Where men are few, and few tools, a few weapons, and their dawns are beautiful. (2: 310)

But at this point, we have so disturbed the natural rhythms that it is impossible to get back to a pre-technological state. We have moved so far away from nature; we have choked the earth by sheer numbers. Many deep ecologists feel that it would be best if the industrial world were just to collapse. Such a collapse is visualized by Jeffers too as the only way out, and therefore the charge of misanthropy is equally strident when applied to both deep ecology and to the poetry of Jeffers. But being anti anthropocentric is different from being misanthropic. What Jeffers, like the deep ecologists, is trying to do is to dislodge the ingrained notion that human beings are the most important species on the planet, rather than that they should be wiped out from the face of the earth.

"Carmel Point" dwells on "the extraordinary patience" of nature that suffers with fortitude the ignominy that man inflicts on it. It knows that its present diminution and defacement are just temporary phases in the cycle of existence: "It knows the people are a tide / That swells and in time will ebb, and all / Their works dissolve" (3: 399). The image of the lost pristine beauty lives on in the very grain of the granite, to be redeemed at the proper time. Jeffers reiterates his concept of time as cyclical; he sees the present deterioration, both cultural and natural, as a sign of a civilization that has gone past the peak, and suggests an ethic that would make human existence on the earth more in tune with the physical world. Freed from his egocentric delusions, having learnt to uncentre his mind from himself and to acknowledge his kinship with the objects of nature, man will acquire the confidence to survive in a world where all his earlier assumptions of centrality have become untenable:

> We must uncenter our minds from ourselves;
> We must unhumanize our views a little, and become confident
> As the rock and ocean that we are made from.
> Where is our consolation? Beautiful beyond belief
> The heights glimmer in the sliding cloud, the great bronze
> gorge-cut sides of the mountain tower up invincibly,
> Not the least hurt by this ribbon of road carved on their
> sea-foot. (2: 523)

Jeffers's concept of time as cyclical assumes significance when his poems are examined from an ecological perspective. Ecology is the science of cyclical patterns, the description of the unending natural cycles of growth and death, and Jeffers's poems reiterate and affirm the cyclical nature of all existence-birth, growth, death, decay, and rebirth. In "Shine, Perishing Republic," the entire Western civilization is moving to the decaying phase of the cycle, settling in the "mould of its vulgarity." However, decay, an inevitable stage in all natural cycles, renders the earth fertile for future growth. "The fruit rots to make earth. / Out of the mother; and through the spring exultance, ripeness and decadence; and home to the mother" (1: 15). While death and decay are part of the natural cycle, and inevitable, what is unnatural seems to be man's rush towards progress and his unnecessary intervention in the natural order of things, upsetting the balance and shortening the natural life span: "You making haste haste on decay". Yet Jeffers affirms life. "Life is good," and all life, however transient, is significant: "Meteors are not needed less than mountains" (1: 15).

Chris Johnston in *Four Dimensions of Deep Ecology* points to two key ideas that have emerged from the call for biological egalitarianism; Systems Thinking and the Gaia Hypothesis. In Systems Thinking, living systems are integrated wholes whose properties cannot be reduced to those of smaller parts. Their essential or systemic properties are properties of the whole, which none of the parts have. Systemic properties are destroyed when a system is dissected into isolated elements. The properties of the parts are not intrinsic properties, but can be understood only within the context of the larger whole (Capra 36-37).

Jeffersian poems like "Natural Music" are attempts to integrate man into the landscape. Man and the landscape belong together and if man has the strength to listen without "Divisions of desire and terror / To the storm of the sick nations, the rage of the hunger-smitten cities" (1: 6), he would find that the still, sad music of humanity is part of the natural scene. It is when man stands apart that he fails to "listen" and loses his "voice." In "Monument," there is no substantial difference between one form of life and another:

> Erase the lines: I pray you not to love classifications:
> The thing is like a river, from source to sea-mouth
> One flowing life. We that have the honor and hardship of being human
> Are one flesh with the beasts, and the beasts with the plants
> One streaming sap, and certainly the plants and algae and the earth they spring from,
> Are one flesh with the stars. The classifications
> Are mostly a kind of memoria technica, use it but don't be fooled.
> It is all truly one life, red blood and tree sap,
> Animal, mineral, sidereal, one stream, one organism, one God.
> (3: 419)

While the parts have value, only the whole is truly important, only the whole endures. "Night" (1: 114-16) has cyclic and cosmic imagery, which moves from man to cosmic whole through the landscape. The central image is here the cosmic whole. But since that whole cannot be comprehended by man, the landscape becomes the metaphor of that whole. "To us the near-hand mountain / Be a measure of height, the tide-worn cliff at the sea-gate a measure of continuance" (1: 115).

The Gaia Hypothesis takes this idea of the shift from the parts to the whole further and applies it to the whole planet. All life on Earth can be seen as a whole that is more than the sum of its parts. Stephen Harding points out in *From Gaia Theory to Deep Ecology*, that the Gaia Theory has revived the ancient view of the Earth as a living organism. Assuming that a life-like quality emerges from the interactions of living beings with each other and with non-living parts of the planetary system, the theory holds that not only does the Earth support individual living organisms and species, but that the sum of all these organisms in the Earth's environment creates a system that is alive. Harding elaborates that such a Gaian approach opens new doors of perception and opens up our vision of the interdependence of all things within the natural world: "There is a symphonic quality to this interconnectedness, a quality which communicates an unspeakable magnificence [...]. As you experience this dynamic, ever-shifting reality, you may suddenly find yourself in a state of meditation, a state in which you lose your sense of separate identity, and become totally engrossed in the life process being contemplated. The contemplated and the contemplator become one."

Poems like "Return," "Vulture" and "Oh Lovely Rock" describe moments of such intense identification with the non-human aspects of nature. "Return" is the recognition of man's spiritual and physical separation from the earth. The poet visualizes a physical immersion in nature, an immersion that is not clouded by the rational human consciousness, which cripples the instinct to participate in the processes of nature. The last two lines of the poem, especially the apostrophe " [...]

noble is the mountain, oh noble / Pico Blanco, steep sea-wave of marble" (2: 409), assumes significance because it reverts to an animistic kind of relationship with the natural world at a point of time, when such a relationship would seem almost an anachronism. But the poet is careful to link this animistic view to a geologic view of the mountain. Pico Blanco when described as a "sea-wave of marble" is related to a time when the rock that constitutes it was molten liquid like the ocean. Connecting the pagan reverence to a modern scientific view of the natural process succeeds in reanimating a sense of wonder in the natural world. The return that the poet attains is a return to a unified organic vision, a recovery of an old emotion based on a modern scientific view of nature. In "Vulture," the poet watches the bird circling above him and inspecting him, as he lies death-still on the hillside. Gradually appreciation of the beauty of the bird evolves into an intense desire to be integrated into nature, a feeling that is almost religious in its intensity: "To be eaten by that beak and become part of him, to share those wings and those eyes—What a sublime end of one's body, what an enskyment; what a life after death" (3: 462).

This heightened awareness of natural objects and an insight that everything is interconnected is a step towards ecological consciousness.. "Oh Lovely Rock" describes the experience of learning to appreciate silence and solitude and rediscovering how to listen. In the first stanza, the reader's eyes are drawn up the rock walls along with those of the poet and the firs that "stare up" the steep rocks, to see forest on forest hung on the mountain ridges. The different trees along with the rock appear to be a single entity.

The second stanza shifts to the human scene; the tenderness and love of the father for his sleeping son seems to be enhanced by the setting. The implication here seems to be that while these human attachments are important, the poet is able to move beyond them. In a moment of attentiveness, the rock wall captures his attention, "as if I were / Seeing rock for the first time." The rock, "light-gray diorite with two or three slanting seams in it," appears to him as living rock, possessing "the silent passion, the deep nobility and childlike loveliness."

The poet experiences a state of being different from his own, yet related to him. The inhumanist vision is attained in such moments of identification with the nonhuman world.

Jeffers's wife testifies to a similar experience in his life as he was handling the boulders during the building of his house: "As he helped the masons shift and place the wind and wave-worn granite," writes Una, " [. . .] he realized some kinship with it and became aware of strengths in himself unknown before" (*Selected Letters* 213). He saw the hardness of his own personality in the boulders and the coldness that would give his "unagitable and somewhat earthfast nature" its characteristic stone tranquillity (1: 112). The kinship that he felt extended the reach of his consciousness and induced something akin to mystical illumination (Karman 48). "Thus at the age of thirty one," continues Una in the same letter, "there came to him a kind of awakening such as adolescents and religious converts are said to experience." Stephen Harding in *What is Deep Ecology*? describes the significance of such an experience:

> A key aspect of these experiences is the perception of gestalts, or networks of relationships. We see that there are no isolated objects, but that objects are nodes in a vast web of relationships. When such deep experience occurs, we feel a strong sense of wide identification with what we are sensing. This identification involves a heightened sense of empathy and an expansion of our concern with non-human life. We realize how dependent we are on the well-being of nature for our own physical and psychological well-being. As a consequence there arises a natural inclination to protect non-human life [...].

From this oneness there arises a deep appreciation of the reality of interdependence, and from this comes the urge to be involved in opposing all sorts of ecological abuses. Such an identification with the physical world is developed in the case of the only two protagonists in the long poems who succeed in attaining an inhumanist vision. The universe imagined as an organic whole echoes the vision of tribal cultures, which in their contact with the earth perceives a spiritual force in nature.

Such a unity with the land will be desirable to Jeffers, a merging of human self and natural other. As Capra says: "Ultimately, deep ecological awareness is spiritual or religious awareness. When the concept of the human spirit is understood as the mode of consciousness in which the individual feels a sense of belonging, of connectedness, to the cosmos as a whole, it becomes clear that ecological awareness is spiritual in its deepest essence" (7). It is such an intuitive recognition of the interrelatedness of all forms, an intense awareness of the natural other that inhumanism calls for:

> . . . I entered the life of the brown forest
> And the great life of the ancient peaks, the patience of stone, I felt the change in the veins
> In the throat of the mountain, a grain in many centuries, we have our time, not yours; and I was the stream
> Draining the mountain wood; and I was the stag drinking; and I was the stars
> Boiling with light, wandering alone, each one the lord of his own summit; and I was the darkness
> Outside the stars, I included them, they were a part of me.
> I was mankind also, a moving lichen
> On the cheek of the round stone. (1: 177)

Extending awareness of other animals and mountains and rivers and stars encourages identification and engenders respect for and solidarity with the field of identification, and Orestes, in "The Tower Beyond Tragedy" calls for such an inhumanist identification with the physical world, in order to attain the philosophical ideal of reaching beyond humanity to a recognition of transhuman reality.

Works Cited

Capra, Fritjof. *The Web of Life: A New Synthesis of Mind and Matter.* London: Flamingo, 1997.

Devall, Bill, and George Sessions. *Deep Ecology*. Salt Lake City: Peregrine Smith, 1985.

Elder, John. *Imagining the Earth: Poetry and the Vision of Nature*. Urbana: U of Illinois P, 1985.

Harding, Stephen. "What is Deep Ecology?" *Schumacher College Web Page.* 28 Jan 2001. <http://schumachercollege.gn.apc.org/articles/stephan2.htm>

____. “From Gaia Theory to Deep Ecology.” *Schumacher College Web Page*. 28 Jan 2001. <http://schumachercollege.gn.apc.org/articles/stephan.htm>

Jeffers, Robinson. *The Collected Poetry of Robinson Jeffers*. Ed. Tim Hunt. 4 vols. Stanford; Stanford UP, 1988, 1989, 1991. (All quotations from Jeffers are from these volumes)

____. *The Double Axe and Other Poems*. New York: Random, 1948.

____. *The Selected Letters of Robinson Jeffers*. Ed. Ann Ridgeway. Baltimore: Johns Hopkins, 1968.

Johnston, Chris. “Four Dimensions of Deep Ecology.” *Hans Hoik Deep Ecology Workshop*. 19 Mar 2001. <http://www.sfsf.com.au/hansfoik_deepquestions.htm>

Karman, James. *Robinson Jeffers: Poet of California*. Brownsville, OR: Story Line, 1995.

Zaller, Robert. *The Cliffs of Solitude: A Reading of Robinson Jeffers*. Cambridge: Cambridge UP, 1983.

5

Descriptions of Landscape in Sylvia Plath's *Crossing the Water* and *Winter Trees*: An Approach*

— B.S. Korde

The word 'landscape' in the title of this paper includes nature in general and also a number of phenomena or manifestations that are parts of nature.

Human being is, as we all know, ultimately, not only a part and parcel of, but also a product of, nature or landscape. This is true to such an extent that human beigs are moulded physically as well as mentally, emotionally by natural atmosphere as they are by their surrounding social, cultural, educational and similar environments. That is the reason why we have so many human races and cultures with their distinctly characteristic features.

It is said that man is known by the company he keeps. This "company" includes social company and environmental company as well. Nature is one of the primal influences on man. A man's physical as well as other related make-up is made by nature itself. The view that nature makes man physically could be substantiated by Darwin's theory of evolution. A number of theories from disciplines such as sociology, psychology, philosophy could be referred to in order to justify that man is influenced by nature / landscape not only physically, but even mentally, emotionally, psychologically etc.

While very convincingly pointing out the relationship between literature and nature (landscape), Patrick D. Murphy writes:

* A pre-revised version of this essay had appeared in the *Journal of Literature, Culture and Media Studies*, No.1 Summer, June 2009. pp.25-36

> Literature and nature have been intimately interlinked in the long history of literary production throughout the world, especially when oral as well as texual forms of literature are taken into account. For example, the ancient Sumerian *Epic of Gilgamesh* records the contradiction between the city and the wilderness and the conflicting values of their ionhabitants. In addition, it builds its plot in part around the destruction of forests through the cutting and transporting of timber to meet urban needs. In China, an entire nature-oriented aesthetic movement, Mountains and Waters, developed and flourished for centuries ... and Dogen's thirteenth-century "Mountains and Waters" sutra influenced not only Japanese poetry in the ensuing centuries but also American poetry in the contemporary period.
>
> While one might say that nature has always permeated literature, it has done so in varying degrees and has been critically received in vastly divergent ways. It has perhaps received no greater neglect in the Anglo-American intellectual tradition than in the early twentieth-century period of modernism and the early years of contemporary postmodernism. While many poets, novelists, and essayists continued to emphasize nature in their writing, the academic establishment tended to ignore such authors and such an emphasis. By the 1960s in the United States, this neglect was initially rectified in the classroom more than in published criticism with courses designed to reflect the rise of the contemporary conservation movement and growing public awareness of a proliferating array of environmental crises.
>
> Through the teaching of courses variously labeled "Man and Nature," "Nature Writing," "Environmental Literature," and "American Pastoralism" as well as the conference presentations and publications that such teaching spawned, a nascent literary canon began to form.[1]

However, although the main premise that we—the human beings—are a part and parcel of landscape remains almost the same for both the Eastern and the Western scholars, it appears that there is some difference also in their attitude to environment/nature. Elaborating on this point, an Indian scholar namely Vidya Niwas Misra states:

> ... The present-day concern of the modern West-oriented man starts with the assumption that we have to take care of the environment as it affects our living. On the other hand, the ancient Indian seer's or poet's (the Seer and the Poet were one) concern started with the assumption that we are a part of a larger existence, in order to become whole, or, in other words, in order to live a full life, we have to offer ourselves to the Universal or to whole or full existence. It is by offering alone that a realization of the whole is achieved and then one's life is fully lived.[2]

Therefore, as nature is taken into account by a number of disciplines while studying human life in its various facets, literature, being a discipline having its roots in human life, also has dealt with the influence of nature on human beings. One of the most important literary examples that shows the influence of nature on human life is John Milton's *Paradise Lost*. Through characters such as the Satan (appearing in the form of a snake) and Eve and through natural phenomena like the Garden of Eden, the Tree and fruit(s) of Knowledge, Milton shows how the whole story of the emergence of human life is shaped and influenced by nature.

The category of writers called 'Regional Writers' is a prominent example of litterateurs dealing with nature / landscape. It is a matter of common knowledge that some of the internationally-known major writers in English in the category of 'Regional Writers' are British writers such as Thomas Hardy (depicting the region of Wessex County) and William Wordsworth (the Lake District), American novelist William Faulkner (Yoknapatawapha) and Indian fictionist R. K. Narayan (Malgudi). These authors show not only various facets of landscape, but they also show how company of landscape shapes human life and character.

In recent times, we have a number of studies dealing with ecological/environmental concerns of literary productions and practitioners of literature—old and new. As far as poetry is concerned, we have some kind of a branch of poetry called nature poetry and a number of Romantic poets have abundantly dealt with nature and its various facets/manifestations including supernatural facets too. Some scholars call literature of and about landscape as "green cultural studies."[3]

As a result of literature written to describe/reflect landscape, environment, nature, or ecology, there has recently emerged a branch of literary criticism called eco-criticism. In this regard, one can refer to books such as Fred Inglis (ed), *Literature and Environment* (1971), Cheryll Glatfelty and Harold Fromm (ed), *The Eco-Criticism Reader: Landmark in Literary Ecology*, Jonathan Bate's *Romantic Ecology: Wordsworth and Environmental Tradition,* John Elder's *Imagining the Earth: Poetry and Vision of Nature*, Karl Kroeber's *Ecological Literary Criticism: Romantic Imagining and the Biology of Mind*, Rueckert William's *Literature and Ecology: An Experiment in Eco-Criticism*, Patrick D. Murphy (ed), *Literature of Nature: An International Sourcebook*, Baidyanath Saraswati (ed), *The Cultural Dimension of Ecology*, and Vidya Niwas Misra (ed) *Creativity and Environment.*

Sylvia Plath had been active as a poet since the era of the 60s onwards, an era which was full of a number of movements in all walks of life—one of which was a movement for ecology. A scholar on Sylvia Plath, namely, Sue Vice, in her attempt to trace the major probable influences on Plath and her writings, one of which is related to ecology, Sue Vice writes:

> To see her work in less biographical terms, the moment of Plath's writing offers a particularly stark combination of discourses, both socially ... and historically—the rise of the ecology and disarmament movements, the legacy of the Second World War and, in the US, the end of the McCarthy era. ...[4]

Further, she writes in the same vein as follws:

> Although Plath's poetry draws upon a variety of twentieth-century events, including the Second World War, the bombing of Hiroshima, the medical disaster of thalidomide and general ecological unease, it is the references to the Holocaust—the Nazi's extermination of the Jews—which have caused the fiercest critical controversy.[5]

Moreover, since in her literary sensibility, Sylvia Plath, it seems, goes quite close to the Romantic poets, landscape is described and reflected her poetry in general and in particular in her collections of poems like *Crossing the Water* and *Winter Trees* chosen for the present study. The reflections of natural phenomena can be seen right from the title of these two collections of her poems—"water" in the earlier collection and "winter" and "trees" in the latter.

As stated at the outset of this paper, the word "landscape" in the title of this paper includes in it nature in general and a number of phenomena that are parts of nature. In the above-mentioned two collections of poems by Sylvia Plath, we can see a number of elements of nature (landscape) described and various facets of theirs presented to the readers by the poet (Sylvia Plath). Such natural phenomena include things such as sea, water, ripples on water, rain, rainbows, snow, the sky, clouds, the sun, sunrise, sunset, midnight, the moon, the stars, light, dark, horizon, the air, wind, mountains, hills, rocks, stones, seasons like winter, birds, and flies like swans, sparrows, bees, flies, animals like cows, sheep, lamb, snakes, jellyfish, different forms of vegetation such as trees, grass, leaves, lilies, blackberries, fruits, reeds, plants, flowers, buds, etc. Moreover, some other natural phenomena like ivory, honeycomb, peasants, marble, deserts, orchids, etc., are also mentioned in her poems chosen for this study.

Plath feels that nature is the ultimate lofty, noble thing. Even the smallest things in the landscape seem to be the loftiest in their appeal. The landscape helps the man to relieve himself of his tensions, pressures etc. Similarly, those cretures who are a part nature possess an innate wisdom of their own. Talking of these things, in her poem "Wuthering Heights", she writes:

> There is no life higher than the grasstops
> Or the hearts of sheep, and the wind
> Pours by like destiny, bending
> Everything in one direction.
> I can feel it trying
> To funnel my heat away.

If I pay the roots of the heather
Too close attention, they will invite me
To whiten my bones among them.

The sheep know where they are,
Browsing in their dirty wool-clouds,
Grey as the weather.[6]

At the same time, the vastness and nobility of the landscape make a sensitive individual aware of his smallness when he compares himself with the loftiness or magnanimity of natural objects. Hence, in her poem "The Surgeon at 2 a.m.", she says:

It is a garden I have to do with—tubers and fruits
Oozing their jammy substances,
A mat of roots. My assistants hook them back.
Stenches and colours assail me.
This is the lung-tree.
These orchids are splendid. They spot and coil like snakes.
The heart is a red-bell-bloom, in distress.
I am so small
In comparison to these organs!
I worm and hack in a purple wilderness. (*CW*, p. 48).

At times, Plath personifies the objects in the landscape. Commenting on some of the natural phenomena, in her poem "Finisterre" she writes:

Now it is only gloomy, a dump of rocks—
Leftover soldiers from old, messy wars.
The sea cannons into their ear, but they don't budge.
Other rocks hide their grudges under the water.
The cliffs are edged with trefoils, stars and bells
Such as fingers might embroider, close to death,
Almost too small for the mists to bother with.
The mists are part of the ancient paraphernalia—
Souls, rolled in the doom-noise of the sea.
They bruise the rocks out of existence, then resurrect them. (*CW*, p. 15).

The sea seems to have been depicted as a disturbing element like warriors who fire cannonballs. Hence, the usages like "The sea cannons into their ear..." or "the doom-noise of the sea".

In Plath's poerty, natural objects are likened to objects and scenes in human, practical life. In short, we get glimpses of different shades and visions of natural objects in her poetry. For example, see the poem "Insomniac":

> The night sky is only a sort of carbon paper, (*CW*, p. 21).

Or

> Nightlong, in the granite yard, invisible cats
> Have been howling like women, or damaged instruments.
> (*CW*, p. 22).

Conversely, she also likens the state of an individual to phenomena of landscape. Here is an example from her poem "Widow":

> Widow, the compassionate trees bend in,
> The trees of loneliness, the trees of mourning.
> They stand like shadows about the green landscape—
> Or even like black holes cut out of it.
> A widow resembles them, a shadow-thing. (*CW*, p. 39).

For Plath, widow is something like a shadow who leads a dark life emotionally. Hence,

> "A widow ... a shadow-thing". (*CW*, p. 60).
> Or in "Whitsun":
> ... the banked rocks sunning in rows, ...

and

> The waves pulse and pulse like hearts. (*CW*, p. 60).

Plath rightly treats even animation as a part of nature.

It is a commonly agreed and experienced fact that landscape is more everlasting than human beings are. Hence, she writes in "I am Vertical":

> Compared with me, a tree immortal
> And a flower-head not tall, but more startling,
> And I want the one's longevity and the other's daring.
> (*CW,* p. 26).

Man can achieve real solace only when he tries to be one with nature. In the same poem she says:

> Then the sky and I are in open conversation,
> And I shall be useful when I lie down finally:
> Then the trees may touch me for once,
> and the flowers have time for me. (*CW*, p. 26).

Due to urbanization and various resultant pressures, in recent times, a great number of human beings have become rough and deceptive—both culturally and emotionally. Most of them are alive merely in the physical sense of the term and not in the human sense of their existence. This is very much in contrast to phenomena in landscape which are either alive or not and nothing in between. In "Stillborn", she writes:

> They are not pigs, they are not even fish,
> Though they have a piggy and a fishy air—
> It would be better if they were alive,
> and that's what they were.
> But they are dead,
> and their mother near dead with distraction,
> And they stupidly stare, and do not speak of her. (*CW*,p.35).

For Plath, articifiality is also a form of death. Therefore, in the same poem, she says:

> They smile and smile and smile and smile at me.
> And still the lungs won't fill and the heart won't start.
> (*CW*, p. 35).

In her poerty, at times, we have purely natural, pleasant descriptions of landscape and its phenomena. Here is one such example in "Private Ground":

The woods creak and ache, and the day forgets itself.
I bend over this drained basin where the small fish
Flex as the mud freezes.
They glitter like eyes, and I collect them all.
Morgue of old logs and old images, the lake
Opens and shuts, accepting them among its reflections.
(*CW*, p. 36).

Pure growth is the hallmark of the natural growth. Such a growth can assume spiritual dimensions too. Resultantly, when an individual senses a spiritual growth, he compares it to some growth in the phenomena of landscape. Plath seems to agree with such a view and hence, in "Love Letter", she writes:

Tree and stone glittered, without shadows.
My finger-length grew lucent as glass.
I started to bud like a March twig:
An arm and a leg, an arm, a leg.
From stone to cloud, so I ascended.
Now I resemble a sort of god
Floating through the air in my soul-shift
Pure as a pane of rice. It's a gift. (*CW*, p. 45).

Right from William Wordsworth onwards, nature poets have expressed their desire to go away from the hubbub and crowds of urban settings as they rightly felt that urban life disturbed a man's peace of mind, unnecessarily filled a man with fatal tensions and worries and hence one wants to have a brerak from all these and get back his pristine peace of mind. The poets and sensitive-minded people have suggested that this peace of mind can be achieved in the company of nature only. In her poem, "Two Campers in Cloud Country", Plath seems to be expressing similar feelings when she writes:

Well, one wearies of the Public Gardens:
one wants a vacation
Where trees and clouds and animals pay no notice;
Away from the labelled elms, the tame tea-roses.
....

Around our tent the old simplicities sough
Sleepily as Lethe, trying to get in.
We'll wake blank-brained as water in the dawn.
(*CW*, pp. 50-51).

("blank-brained", i.e. free from tensions, troublesome thoughts which are a part of urban life).

As we know, once William Wordsworth had said:
The world is too much with us; late and soon,
Getting and spending, we lay waste our powers:
Little we see in Nature that is ours;
We have given our hearts away, a sordid boon!
This Sea that bares her bosom to the moon;
The winds that will be howling at all hours,
And are up-gathered now like sleeping flowers;
For this, for everything, we are out of tune;[7]

It appears Plath has been almost toeing Wordsworth's line in this regard when she says:

Some things of this world are indigestible. (*CW*, p. 58).

At times Plath seems to establish a very close relationship with landscape and its various manifestations so much so that she seems to be treating them as her kith and kin. For example, in her poem "Purdah", she writes:

And should
The moon, my
Indefatigable cousin
Rise, ... (*WT*, p. 7)[8]

It further appears that such a view of hers about landscape is comparable to the view of some people in different parts of the world who have treated natural elements as their kith and kin. One of the example of people who treated nature as their kith and kin is that of an Indian saint-poet from Maharashtra, namely Tukaram who, in one of his poetic compositions, states thus:

The trees and creepers were friends to me,
as also the sylvan beasts;
and the birds spread charm with sweet tunes.
This made the lonely life pleasant,
Besides sparing one from praise and blame.
Sky as canopy, earth as seat,
and scope for roaming where one liked;
Quilt and gourd for bodily needs;
a pleasant breeze from time to time.
Tuka says, one conversed with the Soul;
one discussed with oneself.[9]

The effects of nature on human beings have been fundamental. Such effects have been both pleasant and unpleasant. At one point, to describe both these types of effects of the landscape on the human beings, Plath writes thus in her poem "The Rabbit Catcher":

It was a place of force—
The wind gagging my mouth withmy own blown hair,
Tearing off my voice, and the sea
Blinding me with its lights, the lives of the dead
Unreeling in it, spreading like oil. (*WT*, p. 15)

Plath uses imagery from landscape to indicate pleasant as well as troubled states in the life of a man.

The swans are gone. Still the river
Remembers how white they were.
It strives after them with its lights.
It finds their shapes in a cloud.
What is that bird that cries
With such sorrow in its voice?
I am young as ever, it says. What is it I miss? (*WT*, p. 45)

"Swans" are indicative of innocence, chastity, sinlessness in one's life. It is a stage close to one's childhood—a stage which passes swiftly and one is thrown into the arena of practical life which is full of knowledge—in the sense in which William Blake uses the word in his *Songs of Innocence and Songs of*

Experience—and which, further, is thought to be full of impurity, sin, evils etc.

Plath shows that landscape can fill a human being with the hope of rejuvenation. It shows us the cycle of existence wherein there is death and destruction of something somewhere at one moment and birth and resurrection of something somewhere at the same moment. There is pain and sorrow in the world, but there is also hope of the emergence of something pleasant.

> I wait and ache. I think I have been healing.
> There is a great deal else to do. My hands
> Can stitch lace neatly on to this material.
> The city waits and aches. The little grasses
> Crack through stone, and they are green with life. (*WT*, p. 46)

Plath not only shows that nature is pleasant, beautiful and soothing, but she also, at times, shows the ferocity of landscape/nature. This is a realistic approach to landscape because, it is true that nature is not only constructive and pleasant, but at times, it could also be destructive and unpleasant. Hence, Plath shows the ferocity of a natural phenomenon thus:

> I remember a white, cold wing
> And the great swan, with its terrible look,
> Coming at me, like a castle, from the top of the river.
> There is a snake in swans. (*WT*, p. 35)

These lines by Plath remind us of W. B. Yeats' poem "Leda and the Swan" and background of Helen's life to it:

> A sudden blow: the great wings beating still
> Above the staggering girl, ...[10]

It is in this way that, I think, we can approach Sylvia Plath's poetry to find out her handling of landscape as reflected in her poetry in general and in her above-chosen collections of poems in particular. Of course, the present paper is only one humble attempt in this direction which means I am aware that there could also be better and more elaborate efforts in the interpretation of Plath's poetry from the angle of her descriptions of landscape.

Notes and References

1. Patrick D. Murphy (ed). *Literature of Nature: An International Sourcebook*. London: Fitzroy Dearborn Publishers, 1998; p. xiii.
2. Vidya Niwas Misra. *Creativity and Environment*. New Delhi: Sahitya Akademi, 1992; p.57.
3. Jhan Hochman. "Green Cultural Studies" in Patrick D. Murphy (ed). *Literature of Nature: An International Sourcebook*. London: Fitzroy Dearborn Publishers, 1998; pp. 422-428.
4. Sue Vice. "Sylvia Plath: *Ariel*". In Neil Roberts (ed) *A Companion to Twentieth-Century Poetry*. Oxford: Blackwell Publishers Inc., 2001; p. 500.
5. Ibid., p. 503.
6. Sylvia Plath. *Crossing the Water* [*CW*]. London: Faber and Faber Ltd., 1971; p. 11. All the quotations in the body of this paper from this collection of Plath's poems are to this edition.
7. William Wordsworth. "The world is too much with us". In John O. Hayden (ed). *William Wordsworth Poems, Volume I*. Harmondsworth: Penguin Books Ltd., 1977; pp. 568-69.
8. Sylvia Plath. *Winter Trees* [*WT*]. London: Faber and Faber Ltd., 1971. P.07. All the quotations in the body of this paper from this collection of Plath's poems are to this edition.
9. J. R. Ajgaonkar. *Tukaram: The Most Renowned and Popular Saint-Poet of Maharashtra*. Trans. into English by R. V. Matkari. Bombay: V. Prabha & Co., 1948; pp. 16-17.
10. W. B. Yeats. "Leda and the Swan". In Peter Allt and Russell K. Alspach (ed). *The Variorum Edition of the Poems of W. B. Yeats*. NY: The Macmillan Company, 1957; p. 441.

Works Cited

Baidyanath Saraswati (ed). *The Cultural Dimension of Ecology*. New Delhi: Indira Gandhi National Centre for the Arts, 1998.

Betty Friedan. *The Feminine Mystique*. New York: Norton, 1963.

Caroline King Barnard. *Sylvia Plath*. Boston: Twayne Publishers, 1978.

Charles Newman (ed). *The Art of Sylvia Plath: A Symposium*. Bloomington & London: Indiana University Press, 1963.

Cheryll Glotfelty & Harold Fromm (ed.). *Ecocriticism Reader: Landmarks in Literary Ecology*. Athens, Georgia: Georgia University Press, 1996.

Christina Britzolakis. *Sylvia Plath and the Theatre of Mourning*. Oxford: Oxford University Press, 1999.

Dana Phillips. *The Truth of Ecology: Nature, Culture, and Literature in America*. Oxford: Oxford University Press, 2003.

Edward Butscher (ed). *Sylvia Plath: Method and Madness*. New York: Seabury Press, 1976.

Eileen Aird. *Sylvia Plath: Her Life and Work*. Edinburgh: Oliver and Boyd, 1973.

Gary Lane & Maria Stevens. *Sylvia Plath: A Bibliography*. Metuchen, N. J.: Scarecrow Press, 1978.

Gary Lane. *Sylvia Plath: New Views on the Poetry*. London: Johns Hokins University, 1979.

Glen A. Love. *Practical Ecocriticism: Literature, Biology, and the Environment (Under the Sign of Nature: Explorations in Ecocriticism)*. University of Virginia Press, 2003.

Greg Garrard. *Ecocriticism*. (New Critical Idiom). NY: Routledge, 2004.

Jacqueline Rose. *The Haunting of Sylvia Plath*. Cambridge, MA: Harvard University Press, 1991.

Jon Rosenblatt. *Sylvia Plath: The Poetry of Initiation*. Chapel Hill: The University of North Carolina Press, 1979.

Judith Kroll. *Chapters in a Mythology: The Poetry of Sylvia Plath*. New York: Harper & Row, 1976.

Lawrence Buell. *The Future of Environmental Criticism: Environmental Crisis and Literary Imagination* (Blackwell Manifestos). Oxford: Blackwell Publishers, 2005.

Lawrence Coupe. *The Green Studies Reader: From Romanticism to Ecocriticism*. London: Routledge, 2000.

Linda W. Wagner (ed). *Critical Essays on Sylvia Plath*. New York: G. K. Hall, 1984.

Paul Alexander (ed). *Ariel Ascending: Writings about Sylvia Plath*. New York: Harper & Row, 1985.

Rajani, P., *The Poetry of Sylvia Plath*. Hyderabad, India: Orient Longman Ltd., 2000.

Simpson, L., *A Revolution in Taste: Studies of Dylan Thomas, Allen Ginsbeg, Sylvia Plath and Robert Lowell*. New York: Macmillan, 1978.

6

The Other Wordsworth: A Female Gaze on the Natural World

— Margarita Carretero-González

To Berli, till we meet again.
"L'essentiel est invisible pour les yeux."

The announcement of a couple of visitors' unexpected arrival disrupts a family's peaceful afternoon. Worried about the strangers' identity, the man and his wife hurriedly hide all sorts of books and pamphlets that may give their revolutionary ideas away. An insistent knocking at the door finally forces the man to open it, just in time to see a woman's back. She quickly turns towards him on sensing that the door had been opened. "I was told you were very fine looking", the woman addresses the man unceremoniously. "Indeed you're quite plain, too scrawny. Still, I'll try to see beyond such things, for my brother's sake".

Straightforward, opinionated and considerate over her brother William is how director Julien Temple chooses to introduce Dorothy Wordsworth to Samuel Taylor Coleridge—for such was the man whose family afternoon Dorothy interrupted—and the audience in the film "Pandaemonium." The portrait is further expanded to present a tremendously active, passionate and enthusiastic woman, who finds fault at Coleridge's poetry because of its domesticity, and urges him and her brother to leave the sequestered spot of a library and go outdoors to write "an epic of nature in the language of liberty", the film thus hinting at Dorothy's decisive role in the composition of *Lyrical Ballads*.

Although the film focuses on the character of Coleridge and the jealousy his geniality unwittingly aroused in William Wordsworth, "Pandaemonium" offers a very interesting fictional depiction of the female characters in these men's lives. At the end of the film, an elderly Dorothy turns out to be the only person who completely remembers Coleridge's lost *Kublai Khan*, which she recites, in the midst of her insanity, to a whole audience gathered around her wheelchair. Always according to the film's narrative, it was Dorothy, not her brother William the one who shared Coleridge's visionary powers.

Traditionally, literary history has relegated Dorothy Wordsworth (1771-1855) to the background in the triad she formed with her brother William and S.T. Coleridge. The fame was for William and Samuel whilst Dorothy acted as a force that glued them together while providing her brother with the support and stability he required to keep on writing. On account of all that, Wordsworthian scholars have ransacked her journals almost exclusively for what they had to offer in connection with the poet's works. Readers who approach the entries looking for a sketch of the two great poets at work will be, however, disappointed; they register "no intimate glimpses of Wordsworth and Coleridge about the house, no reports of their literary discussions through the night"(Clark 9). Yet, anyone who reads them in search of their intrinsic value will be most rewarded by Dorothy's extraordinary powers of descriptive detail in rendering everything and everybody around her, most particularly the natural world of her beloved Lake District.

Dorothy's own brother acknowledged the key role her sister played on his development as a poet in the famous verses of *The Prelude*: "She, in the midst of all, preserved me still/A Poet, made me seek beneath that name/My office upon earth, and nowhere else." (Book 10, ll. 919-921). In "The Sparrow's Nest" he records that it was Dorothy who taught him to look, listen ("She gave me eyes, she gave me ears") and, consequently respond to the natural world in a different fashion. In more recent times, the tendency among scholars has been to appreciate their mutual influence. As Virginia Woolf suggested, brother and sister may not have shared the same speech, but they did partake of the same mood, only that Dorothy "stored [it] in prose and later William came and bathed in it and made it into poetry"(Woolf 169). At the same time, Dorothy was most certainly aware of the reading public of her journals: her own brother, so it is also likely that she wanted to store that mood in a way that it would be of use to him. The journals proved useful indeed, not only for William, but also for Coleridge, since both "lifted and used" her phrases and descriptions for their own compositions" (Wilson 5).

Nevertheless, as Wilson remarks, using the journals simply as background material to the works of two of the greatest poets in English literature "directs us away from the startling originality of her [Dorothy's] voice and the strangeness of the story that reveals itself" (Ibid.). Accordingly, this article aims at contributing from an ecocritical perspective to the work being done on revaluing Dorothy Wordsworth's work for its own merit, irrespective of how it contributed to her brother's poetry. If Wordsworth was indebted to her for his perception of nonhuman nature, this article will concentrate on recovering that original inspiring look, the female gaze on the natural world contained within the pages of the journals Dorothy wrote at Alfoxden in 1798 and Grasmere between 1800 and 1803. No study, however, can do justice to the beauty contained within their pages: the changes of the seasons, of the human and nonhuman life around her, or of Dorothy's own mood. They reveal a tremendously acute woman, completely devoted to her brother to the extent of self-effacement. She certainly occupied the proverbial role of the great woman behind a great man, being a woman who had enough talent to shine by herself, not just as the other Wordsworth, if only she had had at her side another Dorothy-like creature to help her in the way she helped her William.

Although the times in which Dorothy Wordsworth lived and wrote relegated her to the domestic sphere on account of her gender, she nonetheless benefited from a degree of freedom unimaginable for many of her female middle-class contemporaries. The life she chose to live with her brother meant, in the words of Virginia Woolf (165-166), that

> they could address themselves undisturbed to the absorbing occupation of living in the heart of Nature and trying, day by day, to read her meaning. They had money enough at last to let them live together without the need of earning a penny. No family duties or professional tasks distracted them. Dorothy could ramble all day on the hills and sit up talking to Coleridge all night without being scolded by her aunt for unwomanly behaviour. The hours were theirs from sunshine to sunset, and could be altered to suit the season [...]. Custom, convention, everything was subordinated to the absorbing, exacting, exhausting task of living in the heart of Nature and writing poetry.

Needless to say, it was Wordsworth's writing of poetry that everything was subordinated to, including Dorothy's own needs. However, among the incessant domestic chores she found enough time to devote to her walks and, fortunately, to her own writing.

Dorothy started keeping a journal at Alfoxden precisely the moment she knew she and William would have to leave the house they had leased in six months' time. Moved by an eagerness to keep everything she was going to lose so soon, she started by describing the prospect the house offered to her, very much as someone of her sensibility in her situation today would take pictures of every beloved spot. She needs no such technological advances; all she needs is words. Many entries are like snapshots, as if written in a hurry, with Dorothy more intent on just recording whatever memory she wanted to keep forever with her rather than on elaborating it. Others, however, constitute what Thomas De Quincey, in his *Recollections of the Lake Poets* defined as "a monument to her power of catching and expressing all the hidden beauties of natural scenery, with a felicity of diction, a truth and strength, that far transcend Gilpin, or professional writers on those subjects" (241). The first entry of the diary, written on 20 January 1798, provides one of the best examples of the merits De Quincey praised:

> The green paths down the hill-sides are channels for streams. The young wheat is streaked by their lines of water running between the ridges, the sheep are gathered together on the slopes. After the wet dark days, the

> country seems more populous. It peoples itself in the sunbeams. The garden, mimic of spring, is gay with flowers. The purple-starred hepatica spreads itself in the sun, and the clustering snow-drops put forth their white heads, at first upright, ribbed with green, and like a rosebud when completely opened, hanging their heads downwards, but slowly lengthening their slender stems. The slanting woods of an unvarying brown, showing the light through the thin net-work of their upper boughs. Upon the highest ride of that round hill covered with planted oaks, the shafts of the trees show in the light like the columns of a ruin. (Wordsworth 2001: 141)

Dorothy always complained that she was not good at drawing or painting, but the above passage proves her wrong: she could paint, indeed, with words. In her hands, the piece of paper becomes a canvas depicting the prospect—which she could have possibly been observing from a window—exquisitely, taking the reader's view from the more distant hillsides to the proximity of her garden, where she concentrates on the colour, shape and movement of the flowers, to take us back to the distant woods. The soul inhabiting every single natural being is suggested in an unpretentious style which, however, attests to "the importance of her 'picture-making power'"(Nabholtz 118).

Although familiar with the picturesque tradition that informs many of her entries, Dorothy was trying to find a voice of her own. In agreement with De Quincey's observations, Lucy Newlyn insists on the fact that "she was articulating an independent aesthetic" (328). Like the painter who is absent from the picture, there is no trace of Dorothy in this first entry of the journal. The sentences are perfectly balanced, written in the present tense, which gives them, following Newlyn (327) "a celebratory mood—almost psalmic", while the rhythm also changes, quickening "as if in response to the awakening countryside".

The second and third entries, however, written on the 21st and 22 January respectively, begin with Dorothy fully immersed in the landscape she depicts: "Walked on the hill-tops—a warm day", and "Walked through the wood to Holford". In Dorothy's brief accounts of both days, the reader perceives the observed landscape, the observer sitting "under the firs in the park", and, occasionally, s/he even shares Dorothy's queries: "Are the male and female flowers on separate trees?" (141)

Quite often, Dorothy seems to have no time to keep the journal up to date, so she writes from memory a string of days at one sitting. On many of these occasions, she writes in a hurry, summarising a whole day in a single line. Thus, on March 28th she "Hung out the linen"; "Coleridge dines with us" on the 29th; on the 30th she "Walked I know not where", and on the 31st she simply "Walked" (Wordsworth 2001: 150). The latter verb is precisely the word that most frequently opens an entry in the Alfoxden journal. She may or may not have read Mary Wollstonecraft's *Vindication of the Rights of Woman*, but Dorothy did not need her advice to women concerning the necessity of taking up exercise in order to keep a healthy, strong mind in a healthy, strong body; she had always been a keen walker. In April of the same year, possibly enjoying the advancing spring, a series of identical entries record a daily walk in the woods. The verb "walk" in its past tense opens twelve of the twenty-seven entries of that month, but the activity of walking is included in every single day except for one, and that just because Dorothy seems unable to recollect what she did that day, judging from the suspension points that follow the date of the 19th. Whether Dorothy "walked", "strolled", "went" or "set forward to" somewhere, alone, with William, Coleridge, or both, by day or by moonlight, the Alfoxden journal presents her always on the move. In her spare time, walking was one of the activities she most enthusiastically embraced; the other was writing.

Dorothy Wordsworth wrote constantly, not only her journals, but also letters, poems, brief narratives or travelogues. The journal, however, was her preferred mode of expression, one at which she excelled. Everything finds its way in the pages of her journals, from the sublime to the mundane, birds' songs rubbing shoulders with bowel problems. Dorothy hoes peas, weeds, longs for a book of botany, irons, reads, saunters in the garden, dyes, starches, plants, copies her brother's poems, tells us about the people around her, suffers from headaches, is happy or in miserable spirits, walks and writes incessantly. In her hands, according to Susan M. Levin (xi) the journal "becomes a powerful narrative of emotional, domestic, and artistic life sustained among a group of extraordinary people." The short journal kept at Alfoxden leaves Dorothy sleeping at Cross, after having walked at Chedder on 17 May 1798, although she miswrote the date as the 22nd. One year and a half later, she began to write another journal in their home at Grasmere. Its pages offer some of the best instances of nature writing in English literature.

The Grasmere journal opens on a very melancholy note. The first entry, very different in tone from the one found on the Alfoxden journal, anticipates a mood which recurrently affected Dorothy at this time of her life. We find her sitting by a lake, after seeing her two brothers off. No longer is she absent from the scene; on the contrary her feelings impregnate all her surroundings. The pathetic fallacy being, according to Frances Wilson, (121) Dorothy's favourite literary device, the lake looks now to her "dull and melancholy, the weltering on the shores seemed a heavy sound"; her state of mind is such that she manages to feel better only "after a flood of tears" (Wordsworth 2001: 1). For the first time since they lived at Dove Cottage, Dorothy had been left alone in the house after William and John had left for Gallow Hill, near Scarborough, on a visit to Mary Hutchinson. Dorothy keeps to herself the reasons why she is not a member of the party—after all, Mary is also her friend—but Wilson (23) is almost certain that the visit is on account of William having decided to marry Mary: "Dorothy and William, who share everything together, have agreed that this visit is something William has to do alone".

If the Alfoxden journal "marked the impending loss of a home" the one kept at Grasmere "celebrated a shared homecoming" (Newlyn 331). It is a much more elaborated journal, kept for a longer and more intense period of time, a time of important changes for Dorothy. It evidences the same frenzy activity engaged in at Alfoxden, but also offers many more instances of the "excessive organic sensibility" pointed out by De Quincey. (237) As a child, that sensibility had made Dorothy afraid of touching a butterfly for fear of removing the dust off its wings and leaving it unable to fly (Wordsworth 2001: 78), or feed robins for private amusement "until they would fly into her room, a pastime that lasted well into her own very old age" (Gittings & Manton 26). As a grown-up woman, this receptivity to the natural world revealed itself, for instance, on one occasion in which she planted again a strawberry blossom she had impulsively uprooted from a rock seconds before:

> [T]he little slender flower had more courage than the green leaves, for they were but half expanded and half grown, but the blossom was spread full out. I uprooted it rashly, & felt as if I had been committing an outrage, so I planted it again—it will have but a stormy life of it, but let it live, if it can. (Wordsworth 2001: 61)

Yet, the same sensibility that made her so receptive to the needs of others also rendered her "perpetually subject to self-conflict" (Wilson 18). The Grasmere journal informs us of frequent headaches, anxiety attacks, miserable spirits and abrupt changes of mood that could bring Dorothy from laughter to tears in a matter of seconds. Five days after having seen her brothers off, she is again thrown into "the deepest melancholy" by "[t]he quietness and still seclusion of the valley" (Wordsworth 2001: 4). There are times when the happiest of prospects prove helpless in lifting up her spirit, or they even bring it further down. After describing "[e]verything green & overflowing with life & the streams making a perpetual song with the thrushes & all little birds, not forgetting the stone chats", Dorothy introduces herself fully on the scene with a short "I grew so sad" (Ibid). Similarly, after a visit from Coleridge, "[t]he sunshine—the green fields and the fair sky made me even sadder; even the little happy sporting lamps seemed but sorrowful to me." (Ibid. 89)

The vignettes of domestic tranquillity usually present Dorothy at her most content. A good example of this is provided in the entry of May 6th, 1802. The use of the present tense and of deictic references together with the erratic punctuation confer the following extract an intimate conversational style that allows the reader to share the depicted scene at the same time as it was being experienced:

> A sweet morning we have put the finishing stroke to our Bower & here we are sitting in the orchard. It is one o clock. We are sitting upon a seat under the wall which I found my Brother Building up when I came to him with his apple [...]. It is a nice cool shady spot. The small Birds are singing—Lambs bleating, Cuckow calling—the Thrush sings by Fits, Thomas Ashburner's axe is going quietly (without passion) in the orchard—Hens are cackling, Flies humming, the women talking together at their doors—Plumb & pear trees are in Blossom, apple trees greenish—the opposite woods green, the crows are cawing. We have heard Ravens. The Ash Trees are in blossom, Birds flying all about us. The stitchwort is coming out, there is one budding Lychnis. The primroses are passing their prime. Celandine violets & wood sorrel for ever more—little geranium & pansies on the wall. (Ibid.96)

The coiness of domesticity, the warm feeling that the perfect interaction of human and nonhuman nature provided her, all this Dorothy richly renders to the reader both visually and aurally.

Frances Wilson (168-169) observes that, in the last of the notebooks that make up the Grasmere journal, Dorothy concentrates almost all her attention on the passing from spring to summer, particularly "what grows and is torn from the earth [...] the birds who migrate and those who stay put". She also establishes a neat distinction between the way Dorothy responds to the ordered nature in the garden when things happen "without a passion", and the sublime beauty of the hills, valleys and lakes beyond the garden. According to Wilson "the beauty is too great [...] there is [...] a grandeur in contrast to which the self feels diminished". In those cases in which the beauty of the valley is overwhelming, "Dorothy wants to express not a scene but an idea; in the excessiveness of the green she is saying something about the power and terror of what lies within". For Dorothy, emotional safety was incompatible with any excess, even if it was an excess of beauty.

There are, however, other feelings to be experienced apart from the safety provided by the harmless environment of the garden. If, following Wilson again (173-174), nature beyond the gate "represents for her a site of awe as opposed to pleasure. The first is a nature with which she identifies [...] and the second is a nature that makes her feel like a different person", this other person finds in the untamed landscape, a seed for many of the sublime descriptions populating the journal. On Sunday 27th of July 1800—which Dorothy again writes wrongly as the 26th—after rowing to Loughrigg Fell, Dorothy and William walk up to view Rydale and spend a long time looking at the lake, "the shores all embrowned with the scorching sun". Walking down by Benson's wood home, the prospect changes:

> The lake was now most still & reflected the beautiful yellow & blue & purple & grey colours of the sky. We heard a strange sound in the Bainriggs wood as we were floating on the water it seemed in the wood, but it must have been above it, for presently we saw a raven very high above us—it called & the Dome of the sky seemed to echoe [sic] the sound—it called again & again as it flew onwards, & the mountains gave back the sound, seeming as if from their center a musical bell-like answering to the birds hoarse voice. We heard both the call of the bird & the echo after we could see him no longer. (Wordsworth 2001: 14)

The call of the raven echoing in the background enables Dorothy to perceive a sort of conversation between bird and mountains that the human beings present in the scene can only look and marvel at. One can even fancy the bird's call fading in the distance anticipating Keats's nightingale's "anthem fad[ing] / Past the near meadows, over the still stream, Up the hill-side". Once more, Dorothy displays her powers of recreating a scene for the reader's eyes and ears.

Dorothy learnt or was born with the capacity to observe, respect and love the natural world for its otherness, irrespective of how it contributed to her own mood. Some journal entries evince that she may have held it sacred, too. Without being a religious woman (at least not in the conventional way; the journals do not record the observance of Sunday Mass, and a walk in the country appears as the perfect way to celebrate Easter), some encounters with nature are presented as instances of spiritual revelation. This quintessential Romantic trait finds its way in Dorothy's writing in her usually simple and unaffected manner when on 3 September 1800 she describes herself "affected to tears" whilst attending the funeral of a fifty-six year-old woman who had "no near kindred, no children". On leaving the dark interior of the house, she is greeted by the delicious scene in front of her:

> When we got out of the dark house the sun was shining & the prospect looked so divinely beautiful as I never saw it. It seemed more sacred than I had ever seen it, & yet more allied to human life. The green fields, neighbours of the churchyard, were green as possible & with the brightness of the sunshine looked quite Gay. I thought she was going to a quiet spot & I could not help weeping very much [my emphasis]. (Ibid. 20)

The harmonious relationship between human and nonhuman nature she had enjoyed in the garden, includes here the divine as a third party, a divinity not to be afraid of but "allied to human life". If she had any religious temperament at all, this was stirred particularly in the presence of nature. At such natural sights, Gittings & Manton affirm, "belief came naturally as breath". (121)

Walking was for Dorothy not only a way of encountering nature, or of sharing a ramble with a loved one; it was also a way of finding her centre whenever she walked alone. The journal written at Alfoxden records many walks with Coleridge; at Grasmere her most frequent companion was William. She undoubtedly enjoyed their company and their conversation, but the way she experienced the natural world must have been different when she walked alone. Indeed, most of the descriptions following those solitary walks are less descriptive and more confessional in tone. Observing the way she includes herself in the pictures her words paint, one may wonder if she perceived herself as one of the fragile creatures she was so fond of protecting, like the butterfly she did not dare touch or the slender, courageous strawberry blossom she planted again to allow it to live its own life.

The way she projected her low spirits on the "dull and melancholy" lake on 14 May 1800 has been pointed out above. If Wilson is right in presuming that Dorothy did not accompany William on her journey on account of his decision to discuss marriage with Mary, it is likely that Dorothy was afflicted by fears of how her brother's new marital status could affect her position in the household. With no one by her side to distract her from the surrounding landscape, she gives free rein to her most intimate feelings, some of which she might not even admit to herself on any other occasion. An attentive reading of the well-known "daffodils" entry may shed some light on this issue if contrasted with an event she recorded two years before.

On Monday 23rd of June 1800, William and Dorothy had gone out on one of their many rambles and were sitting at the foot of a lake, enjoying the wonderful view until Dorothy's head "ached with cold". When William decided to go down to the water to fish, she tried to find repose for her aching head: "I lay under the wind my head pillowed upon a mossy rock & slept about 10 minutes which relieved my headach [my emphasis]" (Wordsworth 2001:13). The daffodils that she admired on 15th April 1802, "grew among the mossy stones about & about them, some rested their heads upon these stones as on a pillow for weariness & the rest tossed & reeled & danced & seemed as if they verily laughed with the wind that blew upon them over the Lake, they looked so gay ever glancing ever changing" (Ibid: 85). Although almost two years separate both entries, it is interesting to note that Dorothy portrays some of the daffodils in exactly the same position she adopted when trying to get rid of her headache; daffodils and Dorothy rested their weary heads upon mossy stones. William's debt to Dorothy's description in "I wondered lonely as a cloud" is patent, but he chose to include only one of the two groups of the flowers she describes in the journa: the tossing, reeling, dancing, happy majority, "fluttering and dancing in the breeze", while he excluded those which, like Dorothy, took the stones as their pillows. Taking this resemblance into account, one might speculate about whether Dorothy perceived herself as the small group of weary daffodils, unable to laugh and dance like the bigger group. Did she feel she was being left out?

The toll that William's decision to marry took on Dorothy has been explored in detail by Frances Wilson; it is, however, an issue on which all biographers seem to agree and which a look at the parts of the journal written in the months previous to the wedding corroborates. William was marrying Mary

Hutchinson the following October and had planned to travel to France that summer with Dorothy to visit Annette Vallon and the daughter they had in common. Dorothy felt that a whole cycle was closing and she took pity on the position Annette and Anne-Caroline would be relegated to after the nuptials. The tension she suffered at that time proved to be too much to bear and the journal records her inability to sleep, recurring headaches and a general state of low spirits; even her beloved garden goes "overrun with weeds." (Wordsworth 2001: 117)

Fond as Dorothy was of observing the songs of birds, her entry of Saturday 19th of June describes "our own Thrush shouting with an impatient shout—so it sounded to me. The morning was still, the twittering of the little Birds was very gloomy [my emphasis]." (Ibid: 112) Her emotional response to the collapse of a swallows' nest the progress of which she had been observing for ten days is indicative of how sensitive she was not only to the birds discomfort, but also to her own situation facing the approaching changes: "Poor little creatures", she writes, "they could not themselves be more distressed than I was." (Ibid: 115) Fortunately, the swallows managed to complete their nest on July 6th, just in time to add a merry note to her impending journey to France two days later. She might very well have taken the successful reconstruction of the nest as reassuring, a good sign of the continuation of her happy life with her brother even with Mary in the picture.

And yet, she knew things could not be the same, that she herself would change. Awareness of the way "attachment for Dorothy was a continuation of selfhood" (Wilson 32), renders her leave-taking of Dove Cottage before the journey to France the more poignant since we realise that in saying goodbye to every beloved object and creature Dorothy is parting from one side of herself that will never come back once Mary becomes Wordsworth's wife: "I must prepare to go—The Swallows I must leave them the well the garden the Roses all—Dear creatures! They sang last night after I was in bed—seemed to be singing to one another, just before they settled to rest for the night. Well I must go—Farewell. ..." (Wordsworth 2001: 119)

Dorothy's life certainly changed, although not as she feared. The family group now included Mary and the children she had with William, and Dorothy was ever a key member in the household. Together they shared a life full of happy and tragic moments, as Dorothy rejoiced on the birth of her nephews and nieces but also endured the deaths of some of them. The new life seems to have taken its toll on Dorothy's writing, for the Grasmere journal ends abruptly three months after the wedding. She kept on writing letters and poems, but they lack the degree of intimacy and spontaneity of the journals: they are for others to read, not for her brother, as the journals were. Dorothy may have felt that once Mary had taken up her role as William's most direct confidante, there was no use in keeping a journal from which nobody would profit. In that belief, once again, her self-effacement proved her wrong. The Alfoxden and Grasmere journals reveal Dorothy Wordsworth as one of the finest nature writers in English literature and it remains a source of pleasure for many more readers than just her beloved William.

Works Cited

Clark, C. (ed.). 1986 [1960]. "Introduction" to D. & W. Wordsworth, *Home at Grasmere. Extracts from the Journal of Dorothy Wordsworth and from the Poems of William Wordsworth*. Harmondsworth: Penguin Books.

De Quincey, T. 2009 [1839]."From *Recollections of the Lake Poets*" in S.M. Levin (ed.). *Dorothy Wordsworth*. New York, London, etc.: Pearson Education & Longman: 237-242.

Gittings, R. & J. Manton. 1985. *Dorothy Wordsworth*. Oxford: Clarendon Press.

Levin, S.M. (ed.). 2009. *Dorothy Wordsworth*. New York, London, etc.: Pearson Education & Longman. I read Susan M. Levin's book in the summer of 2008 and this is the year in which I am also writing this article. For some reason, however, the book is copyrighted 2009 and so appears in this list of works cited.

Nabholtz, J.R. 1964. "Dorothy Wordsworth and the Picturesque", *Studies in Romanticism*, vol. 3, no. 2: 118-128.

Newlyn, L. 2007. "Dorothy Wordsworth's Experimental Style", *Essays in Criticism*, vol. 57, no. 4: 325-349.

Temple, J. (Dir.). 2002. *Pandaemonium*. London: BBC Films.

Wilson, F. 2008. *The Ballad of Dorothy Wordsworth*. London: Faber & Faber.

Woof, P. (ed.) 2001. "Introduction" to D. Wordsworth, *The Grasmere and Alfoxden Journals*. Oxford: Oxford University Press: ix-xxi.

Woolf, V. 1986 [1932]. *The Second Common Reader*. San Diego, New York & London: A Harvest Book, Harcourt, Inc.

Wordsworth, D. 2001. *The Grasmere and Alfoxden Journals*. Edited by Pamela Woof. Oxford: Oxford University Press.

Wordsworth, D. & W. *Home at Grasmere. Extracts from the Journal of Dorothy Wordsworth and from the Poems of William Wordsworth*. Edited by Colette Clark. Harmondsworth: Penguin Books.

Wordsworth, W. 1994. *The Works of William Wordsworth*. Ware, Hertfordshire: Wordsworth Editions Ltd.

7

Embodying Nature through Aesthetic Experience

— **Priyadarshi Patnaik**

What you see is not what you experience

Exploring the correlation between nature and aesthetics can primarily be done in two distinctive ways:

i) experiencing aesthetic qualities in nature, and

ii) internalizing nature in an aesthetic form and then appreciating it.

Experiencing nature as a work of art has been problematized in the twentieth century and yielded a form of aesthetics which can be broadly called "environmental aesthetics."[1] Traditionally, aesthetic form presupposes an "author," or creator to whose consciousness the aesthetic act can be attributed. In this schema, the movement is from "author" to "art" to "appreciator." It also brings in the notion of "intentionality." Traditionally, an act of art is a conscious and intentional one by its author. For instance, a painter decides what to paint, how and has a reason why s/he performs the act. The work of art and the author are thus linked. The work has a father (mother) or point of origin and in social practice, the author is responsible for the work. Even when a work is anonymous it is a work of human hand and thus still has an author (whose name is unknown). Even when aesthetic stands relegate the author to the background and explore art as an autonomous entity, the function of the artist as author is acknowledge, even if it is thrown to the background. Looking at nature as an aesthetic entity challenges this notion of a creator or author behind the act. Sometimes, it cannot even be called an "act." The clouds in a landscape might have formed yesterday while the trees are a thousand years old and the mountain a million years before that. Looking at God's act as sometimes an aesthetic one undermines his creativity. Intentionality and aesthetic reformulation finally rest in the hands of the perceiver, viewer and critic. The act of aesthetic appreciation of nature becomes the moment of both creation and appreciation of the object under scrutiny at the same time. This is a strand which has been explored in the West in recent times, diverse perspectives about how one perceives nature are presented and different theories of aesthetics of nature have emerged.

However, an Indian perspective would have to look the other way, at the way nature has been internalized into aesthetics; more interestingly, how nature has been embodied through art.

Aestheticizing art involves an act of framing. Let me give an example from painting or photography—visual arts. All around us are objects and things that have the potential to be transformed into art. But the artist has to make a choice of what to show and how. When nature is represented, say, through photography, what is put within the frame is at the cost of the rest of the world that is excluded from the frame. Only that which is contained within the frame forms the aesthetic object.

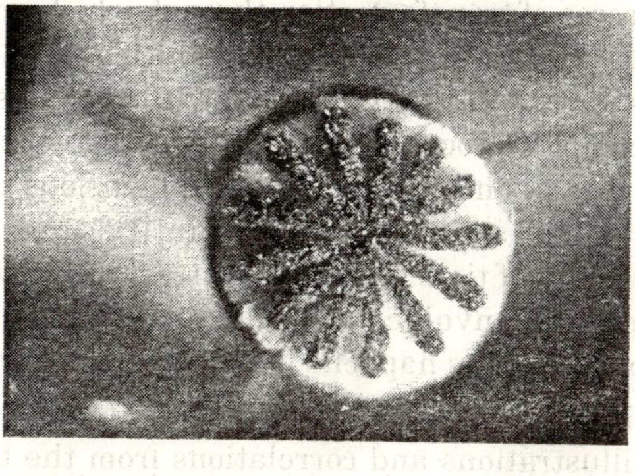

***After the Bloom*. Photograph by author. Lumix FZ50 with macro attachment, 2009. By eliminating the other flowers from the frame the bud is decontextualized and then recontextualized in almost abstract terms in relation to its frame and background colours**

Although it doesn't concern us directly, in appreciating nature as aesthetic object we also use frame—a conceptual frame of aesthetic canons and a perceptual frame of what to look at and how to look. What I referred to a little earlier in the analogy of the photograph constantly happens in the naturalistic tradition of painting as in figurative art or photography. The artist/photographer decides what to put within the fame and how. Here, what one sees is what one experiences. This is not to say that transformations do not take place. Transformation of natural objects in representational art takes place in many ways in Western art tradition—Impressionism, Imagism, Symbolism, Expressionism and Surrealism are a few instances where the world outside is transformed.

But when we talk of embodying nature in Indian art, we are talking about something distinctively different. While I have no intention here of comparing different aspects of Indian art with Chinese art or different periods of European art, I wish to insist that if we take as our case-study a distinctive form of painting —the *ragāmālā* tradition—we cannot use the notion of framing used above, nor can we talk of transformation as is explored in many European art movements; rather we see the embodiment of nature in a miniature scale. Aesthetic canons do operate, but they come later. What happens first is a complex process of transformation through which art manages to evoke and then embody nature—so much so that art can be explored as nature in the miniature by the connoisseur. This involves a complex process of meaning-making which precedes aesthetics. It also encompasses other dimensions of art (and tradition) such as music, spirituality and the very conceptualization of time through nature.[2] True, any process of meaning-making involves the socio-cultural context and language, but here this happens in a much more involved way.

In this brief essay I shall attempt to analyze how it is done with some illustrations and correlations from the tradition of music and painting, in the background of meaning-making strategies that evolve out of meditative tradition and rules of code of conduct of everyday life. It is in this sense that the miniature art tradition called *ragāmālā* leads to a position where what you see is not what you experience, because the experience comes through a series of mediations after the act of seeing—acts which lead to meaning making—and that finally embody nature in art. It is at this point that aesthetic canons merge in the process and the successful work of art embodies nature and the poor one fails.

Before we can talk of *ragāmālā* paintings we have to talk of music, of spirituality and of cultural understanding of and codification of time. This is so because these paintings absorb and internalize certain aspects of these practices and the very act of embodiment of nature in art can only be experienced by the ideal perceiver (*sahrudaya*) with knowledge of the cultural codes through which music and contemplation of music (*dhyāna*) take place. Social practices of mapping one's life according to the configurations of the spatial world—which one must travel and touch at the appropriate moment—defines time for the Indian. If time is a flow and invisible it can only be grasped by the changes that take place around us. We do not see time but the acts of time where seasons change and the day becomes night. Doing things at the "appropriate time"[3] (*shubha muhurta*)—which is a key concept in Indian cultural life—can be made possible only by observing the various markers of time—stars, sun and moon, seasons, leaves, the texture of breeze—only through which time can be grasped and experienced. The reason for bringing in a discussion of time is because markers of time are abundant in nature and experiencing nature involves knowing how time or the cycle of time has changed/influenced and transformed nature. Internalizing nature in art would involve using popular markers of time. Nature can be evoked through markers (say, trees, flowers, clouds) which also express nature at a specific time. Time evoked thus has (or should have) compatibility with the context within which the work of art evolves. Thus, experiencing nature in the miniature would also involve evoking the notion of appropriateness of time: hence, this discussion on time.

In the Indian tradition, almost everything is encased within the notion of appropriateness, as is illustrated above. *Nitishāstras* tell us that there is an appropriate time for everything. Our life is divided into appropriate phases when we should spend our formative years of learning (*brahmācharya*), become householders (*grahastya*), prepare for old age (*vanaprastha*), and renunciation (*sanyāsa*)—*chatur-āshramas*. We have an appropriate time for study, for sleep, for worship, for journey, for marriage, for love-making and even for birth and death. In this tradition each musical note also has its appropriate emotion, its appropriate resemblance to or origin from different animals and birds. Music also has affinity to nature and time, and hence different organizations of note structures (known as *rāgas*) have appropriateness or affinity to appropriate time often depicted through times of days and seasons.

The point I wish to make is that time in Indian music (not musical time of rhythms and beats) is often experienced through seasons.[4] We have *rāgas* like *Hindolā*, *Megh*, *Varshā*, *Bāhār*, *Dipak* and so on which function in two distinctive ways:

i) their vibrations are the essence, the evocation of those seasons; they emerge from those seasons. In the process they acquire and internalize the properties of those seasons;

ii) hence, logically, it follows that at any point of time, when sung, they can "evoke" the nuances of those seasons. In the mind of the ideal listener, they can create a season out of season. But this does not embody the season. The season still lacks a visible body.

The other thing that happens in the tradition of Indian music is the "embodiment" of that "musical form" or *rāga*. In a tradition that is traced back to the third century AD (though the date is a little controversial) Walter Kaufmann points to musical treatises that talk of *dhyānashlokas*[5] or "verses for contemplation." By the seventh century the word *rāga* appears in musical treatises (Matanga's work) and specific seasons and time for each *rāga* is prescribed by the twelfth century in the works of Someswara. In *Ragāsāgara* each *rāga* is linked to a *dhyānashloka*.[6] The *dhyānas* emerge out of a total realization of a musical form so that the *rāga* is embodied, usually as a god or goddess, in a human form. A practicing musician has these as guidelines. In singing a *rāga* to perfection, s/he is able evoke this image in his/her mind. This strategy of meditating on an image with such intensity as to actualize it comes from our spiritual tradition. The notion of ragamala or garland of *rāgas* also goes back only a few centuries. As we have already discussed, seasons can be evoked through music. The tradition of *rāgamālā* is one where, through an unbroken recital of *rāgas*, one after the other, the entire cycle of seasons is evoked. But even here the seasons are not embodied. I wish to point out that it has nothing to do with the medium, for as discussed earlier, music does manage to embody *rāgas* through *dhyānas* or meditative states.

We shall now focus on a different form of art altogether—painting. Our objective, right from the beginning of the paper, has been to reach this point. But the digressions were necessary in order to understand how the diverse traditions and codes from cultural life, music and meditative tradition (*dhyānas*) function in painting to embody nature.

Indian art has non-naturalistic paradigms. The Western tradition of creating an illusion—life-like—which obsessed it from the Greek times never affected Indian art. Hence, it had no difficulty in finding an amalgamation with Muslim painting. Both, at the point of their confluence, had rich traditions of miniature paintings behind them. If book illustration is found in the Muslim tradition, palm-leaf illustrations are found in the Jaina tradition. Art, in such a context, evoked, it did not 'represent." Evocation was through suggestions, through cultural codes, through quintessential qualities that could bring back memories of the whole.[7] Almost a thousand years after the *dhyānashlokās* and the singing of *rāgamālā*, by some unknown accident, in the sixteenth century, the *rāgamālā* painting tradition evolved in the creative ferment where Hindu courts and Muslim courts encouraged art, Hindu artists and Persian artists commingled.

The best of *rāgamālā* paintings internalize music and meditation. But they also demand a lot from their connoisseurs—musical and meditative/contemplative training. At the centre of the painting is the image of the *rāga*, given human/godly form, which evokes time through seasons. This is supplemented by the various signs of nature—a specific type of cloud, a particular colour of night, a seasonal bird or blossom.

The illustrations below, available on Wikipedia,[8] depict *Rāgini Bhairavi* and *Raga Bhairavi* respectively. Pre-dawn and dawn are suggested through dark nights. But blooming flowers and active birds suggest that it is not night, in the first painting. In the second painting, the peacock suggests dawn. In the first image, the female figure is goddess Bhairavi worshiping a *linga* while in the second, Shiva himself is presented.

In the mind of the connoisseur what operates is a certain kind of cadence. Music vibrates and evokes season. The colours and shapes of familiar objects bring back the season. Texts or *dhyānas* written besides the paintings complete the experience. In a space of less than a foot square nature is embodied—the connoisseur, through text, cadence and image experiences the season and all that happens within his/her aesthetic relish.

The aesthetic experience which distinctively involves sensitivity to the subtle nuances of nature and its deep appreciation is achieved here through a process of internalization of certain cultural codes, which when correctly evoked bring about not only an aesthetic experience, but almost bring back nature to be experienced in all its immediacy. It is thus that nature is embodied.

Notes and References

1. For a detailed discussion of the various perspectives in environmental aesthetics, see Allen Carlson, *Aesthetics and the Environment: the Appreciation of Nature, Art and Architecture*, (Routledge and Kegan Paul: London & New York, 2000).
2. For a detailed discussion of *Rāgamālās* and the interrelationship between music and painting, aesthetic emotions and experience see Walter Kaufmann, "*Rasa*, *Rāga-Mālā* and Performance Times in North Indian *Rāgas*," *Ethnomusicology*, Vol. 9, No. 3 (Sep., 1965), pp. 272-291. Kaufmann writes, " *Rasas*, pictorial representations of *ragas*, and their performance times constitute interrelated fields which require some consideration because, in spite of their ambiguities, they have survived the long span of nearly fifteen centuries and did not disappear even when the Indian drama discarded *rasas* and related features." p.272.
3. Indian astronomy and astrology are very closely linked. Since it is believed that astral bodies influence human beings and the environment, and environment influences the fates of humans (indirectly the astral bodies bring this about) a whole science of calculations emerged out of it–*muhurtasastra* or astrology that specializes in identifying auspicious and inauspicious moments.
4. See Swami Prajnanananda, *A History of Indian Music* (Kolkata, Ramakrishna Vedanta Matha, 1963).
5. "In the Indian music tradition, right from the time of *Rāgasāgara* (approximately 2nd century AD) *dhyānas* are to be found associated with specific *rāgas*. Thus, here we find the personification of a *raga* with attributes as well as ambience spelt out. The *rāga* which may internalize, say, spring, through the *solka,* manifests Spring visually in the form of a personification as well as associated ambience – trees in blossom, etc. From this point it is but one step to paint these deities and we have *rāgāmālā* paintings." See Priyadarshi Patnaik, "Time and the *Rāgāmālā* Tradition: A Brief Exploration," *Time in Indian Indian Culture: Diverse Perspectives*, eds, P. Patnaik, S. C. Chatterjee and D. Suar (D. K. Printworld) (Forthcoming 2009). Also see Kaufmann; and Swami Prajnanananda.
6. It's date is uncertain, it being ascribed to Dattila (2nd century AD) and Narada. See Kaufmann, 275-77.
7. It's date is uncertain, it being ascribed to Dattila (2nd century AD) and Narada (12th century AD). See Kaufmann, 275-77.
8. An entire tradition of literary criticism evolved around the notion of "suggestion" or "evocation." It was called the theory of *Dhvani* by the 10th century aesthetician Anandavardana. Later, in the works of Abhinavagupta (11th century AD) *rasas* (aesthetic emotions) were correlated with *dhvani*. Since emotions cannot be shown directly, they can only be evoked through suggestion. This

notion definitely pervades painting in a non-naturalistic tradition where suggestions can be both symbolic as well as emotive.

9. <http://en.wikipedia.org/wiki/ragamala.htm>

Works Cited

Carlson, Allen. *Aesthetics and the Environment: the Appreciation of Nature, Art and Architecture*. Routledge and Kegan Paul: London & New York, 2000.

Kaufmann, Walter. "*Rasa*, *Rāga-Mālā* and Performance Times in North Indian *Rāgas*." *Ethnomusicology*, Vol. 9, No. 3 (Sep., 1965). pp. 272-291.

Patnaik, Priyadarshi. "Time and the *Rāgamālā* Tradition: A Brief Exploration." *Time in Indian Indian Culture: Diverse Perspectives*. Eds, P. Patnaik, S. C. Chatterjee and D. Suar. Delhi: D. K. Printworld (Forthcoming 2009).

Swami Prajnanananda. *A History of Indian Music*. Kolkata: Ramakrishna Vedanta Matha, 1963.

"*Ragamala* Paintings." *Wikepedia*. <http://en.wikipedia.org/wiki/ragamala.htm>

8

People and Forests: An Eco-Social Reading of Kadamanitta's Poems

— P. Rohith

Unlike the eco-romantics,[1] deep ecologists,[2] spiritual ecologists and eco-feminists,[3] who try to remedy the ecological crisis through spiritual, biocentric and romantic means, the social ecologist and eco-Marxists try to root the current ecological crisis in the capitalistic system. Social ecologists conceive over-consumption, productivism and consumerism as symptoms, the causes of which lie deep in the existing social structure characterized by hierarchy, exploitation and domination. Social ecology can be described as a comprehensive way of understanding scientific ecology and social/political systems. It is in this platform provided by the social ecologists that most of the Green parties operate. To the advocates of social/Marxist ecology, it is not the magnitude of population, but the way in which they relate to one another that has caused the present economic as well as ecological crisis.

Murray Bookchin, one of the major figures among the social ecologists, explains current ecological crisis as rooted in the relation of domination between peop'e, which assumes catastrophic proportions under capitalism.(63) Bookchin traces the human domination of nature back to the disintegration of primitive tribal communities, which were characterized by equal distribution of power and resources. The social evolution that occurred in these communities forced them to move towards a patriarchal system with men assuming complete domination over women. This domination was gradually consolidated by extending the influence of power to the weaker sections of the community. The domination of nature is a natural consequence of this social structure. And so the social ecologists often reject hierarchy, domination, centralism and capitalism.

In the environmental discourse, nature is sometimes understood as a separate entity outside history and human society. However, this essentially romantic notion of nature that is pristine and that is outside history and human context does not have the potential to address the contemporary ecological crisis.[4] O'Connor rejects this notion of nature when he remarks that human history and natural history are "dialectically interconnected" (Darier, 936). Nature, thus is not something to be cherished for its own sake, but acquires value for us only in relation to our interactions with it. Rather than concentrating on human-nature dichotomy, social ecology comprehends all social as well as environmental problems as rooted in the "material fact of inequality" (Veak 403). A critique of prevailing socio-political and anti-ecological attitudes, social ecology presents a reconstructive, ecological, communitarian, and ethical approach to society.

Another stream of thought that is akin to, yet different from social ecology, is ecological Marxism or ecological socialism. The major contributor to this stream of thought is James O'Connor. Different from the social ecologists, eco-Marxists believe that social causes alone are insufficient in explaining economic as well as ecological crisis. From an eco-Marxist perspective almost all problems the humanity and nature face are essentially linked to the political economy. According to O'Connor, "to understand the current environmental crisis it is vital to understand the broader economic and political conditions which created that crisis in the first place" (Darier, 935). It is only through a just distribution of productive resources and capital that this problem can be settled.

In this chapter, I shall attempt to uncover the ecological consciousness reflected in the poems of Ramakrishna Paniker (1935-2008), popularly known as Kadamanitta Ramakrishnan or simply Kadamanitta,[5] written during the seventies and eighties of the last century in the immediate context of the campaign to resist the Silent Valley Hydroelectric Project (henceforth SVHP). The proposed SVHP was to submerge almost a thousand hectares of biologically rich and ecologically significant tropical rainforests and grasslands in the Silent Valley. The Silent Valley is a remote and secluded forest tucked away in the Western Ghats in Palghat district of northern Kerala. The name Silent Valley was probably given by the colonial British officers.[6] Situated 45 kilometres Northwest of Mannarghat, the Silent Valley covers about 89.52 square kilometres and is rich in both flora and fauna. Ecologists describe the Silent Valley as the sole surviving bit of evergreen forest in the Sahya ranges. In the words of the renowned ornithologist, Salim Ali "it is a very fine example of one of the richest, most threatened and least studied habitats on earth" (qtd. in Sugathakumari, *Silent Valley* 11). The fact that this area is devoid of any human habitation makes it a rare ecosystem. The area has an "uninterrupted evolutionary history of almost fifty million years" (Swaminathan, *Silent Valley* 2; Sugathakumari, *Silent Valley* 11; KSSP, *The Silent Valley* 1-2). The Silent Valley is often described as "a botanist's bouquet, an entomologist's enigma and a biologist's delight" (Jayaram 54). There is not even a tribal village in the valley and it had always been difficult to reach. It has therefore remained a well-preserved forest with few incursions. "Dark, cool and vibrating with life the Silent Valley, the richest expression of life on earth," as Sugathakumari remarks, "presents a text book version of the tropical evergreen forest" (Silent Valley 11).

The fraternity that evolved among Malayali writers in order to challenge the SVHP was unique, as such an organised literary/artistic intervention for ecological and social causes was unheard of in the socio-political and literary history of Kerala and even of India. Kadamanitta was one of the most vocal poets among the writer-activists[7] of the movement to protect the Silent Valley. The picture that the Silent Valley forest conjures up is rightly that of a thick, dense, dimly lighted tract of forest with no visible human impact both in the form of settlement and commercial as well as industrial interests. In this aspect, the movement to protect the Silent Valley from the proposed hydroelectric project differs immensely from that of other environmental struggles like Narmada, Chipko, Mavoor and Baliyapal which evidently rely on the rhetoric of social justice and humanitarianism. Interestingly enough, the conspicuous absence of the human element was not a deterring factor that distracted literary interest. Writers' interest in the Silent Valley was not restricted to the ecological significance of the area, rather they were drawn to it for its latent eco-social critiquing possibilities. This position might appear paradoxical for the protection of the Silent Valley employed, directly or otherwise, no vocabulary of philanthropy and local/regional justice. In this chapter, I will read poems of Kadamanitta written/recited between 1975 and 1985 against this backdrop.

Kadamanitta's poems written during the course of the campaign to resist the SVHP are to a great extent influenced and shaped by Marxist and Naxalite ideologies. It is not the sheer depletion of the nature, but its effect on the lives of the common folk that worries the poet. Poems such as "Kiratavrutham" (The Hunter's Tale), "Kattaalan" (The Savage) "Kunhe, Mulappal Kudikkarutu" (Child, Do not Drink Breast Milk) and "Kurathi," all written during this period, depict the intrusion of the modern, capitalist, market systems into the natural habitats of the marginalized. Alongside this recognition, the poems of Kadamanitta depict the loss of the natural, rural simplicity and purity to the industrial, urban, market rapaciousness and ruthlessness. Poems of Kadamanitta try to capture these varied aspects of "development" and the subsequent despoliation of the environment. "Kiratavrutham" vividly portrays this aspect of the contemporary social structure. The protagonist, Kattaalan "with a burning torch in his chest", (8) is furious at the marginalising forces that corrupt and destroy his indigenous lifestyle by imposing urban, market values into his knowledge system. The underprivileged are deprived of their natural surroundings:

> Where is the lightning, thundering sky
> In which I sowed my dreams?
> Where are my tulsi forests and
> Twilights combing their moist lock?
> Where are the green meadows
> In which grasshoppers hop? (36-41)

The Kattaalan's burning torch in his chest alludes not merely to the burning torch pictorically marked on the youth-wing left's flag but to the raging public sphere. Similarly, the images of the "eyes that reflect a breeding tigress" (1-2) and "half-curled eyebrows that sprout on the tail of a black cobra" (3-4) represent the rage of the public including that of the literary figures. Kadamanitta brings it home to his audiences that the endeavour they are collectively embarking upon is nothing less than a war. The abundance of images in "Kiratavrutham" of "bow" and "arrow", "trumpet" and "the stone axe" vividly suggest a battlefield. The same ire informs his other poems as well. In "Kurathi", one of his best read works, the Kurathi is portrayed as "valiance of the burned soil", "violent river", "forest fire" and "chiselled wild-stone". The protagonist of "Kiratavrutham" proclaims:

> The hands of huntsman,
> I'll chop them with my axe.
> They who ruin the mountains
> Must flow headless in the river.
> They who hew the trees,
> They who destroyed my clan,
> I'll garland the earth
> With their entrails. (78-85)

The realization that the poor have been in chains for ages marks the poems of Kadamanitta out from other nature poems written during the period. The comprehension of how the oppressed have been "utilised," "sacrificed," "dominated" and "denaturalized" for the benefits of the ruling affluent minority makes the socio-political side of the ecological crisis evident. Concerns regarding human casualties were admittedly absent from the Silent Valley controversy, though such concerns have been raised against all major "developmental" projects before and after. The poet presents an environmental Naxalite view of the ecological crisis. The following lines from "Kunhe" suggest that the current ecological problem lies in the relation of domination between people:

> They smashed the gratification of
> Your life, sprouted in the hill,
> They dishonoured our clan
> By blasting the atom bomb of knowledge,

They sowed calamities
By burning the paddy fields,
They sacrificed us
To fix the bridge tearing the river,
Now they call us
To weaken the earth.
Now they approach us
To keep us in chains. (31-42)

What is foregrounded here is the realisation that the so called progressive process, in actuality, is detrimental to communities in the fringes. Such processes often drain practical knowledge and native wisdom from indigenous cultures by means of alienating them from their traditional habitats. The significance of this process in relation to small tribes is the subsequent loss of various specific skills and the consequent loss of skilled practitioners to use them. This loss of specific skills would, eventually, drive such communities to urban labour markets for their livelihood. Instances like these reveal Malayali writers' awareness that modern science and technologies cannot adequately substitute the loss of indigenous experiential knowledge.

The revolutionary spirit of Marxism and the Naxalite movement is subtly directed towards the injustices against nature. The declaration, "the onus of today is to chop the hands that strangle Her" takes the revolutionary zeal in Malayalam poetry forward from Idasseri's "we must reap power first, let the ponnaryan[8] be after" ("Puthan Kalavum Arivaalum") to new aesthetic-revolutionary heights. Here one can have glimpses of the Marxist in Kadamanitta, for, as the Left who rely on distributive value, his characters too rebel for political and economic systems where domination and oppression no longer exist. Even then they are aware of the need for establishing ecological justice, for they realise that only the principle of distributive justice has the potential to sustain the environment and far more importantly save the so called communities in the fringes. In other words, as Burgmann, Kadamanitta reminds us "[...] that socialism cannot be built on rubbish heaps, that ecological problems cannot wait until the revolution" (qtd. in Jagtenberg and McKie 116).

Such instances of resentment in these writings attracted the public. Kadamanitta seized this occasion and utilized it to present his concerns which also come up in such compositions as "Kozhi" (The Hen), "Karuthamakkal" (The Black Children), "Uzhavuchalukal Keerunnavan" (The Ploughman), "Adimakal" (Slaves), "Kochumanushyanmar" (Poor Folk), "Harijanangal" (The Harijans), "Daridra Daivangal" (Poor Gods) and "Shantha." The rage of the Kattaalan against the corrupting, modern, capitalist, market forces is not a prayer for reviving tradition; on the other hand it represents his yearning for a just society.

In spite of the relatively just and self-sufficient traditional nature of the tribal societies, in actuality such communities are highly vulnerable in relation to the developed, civilized and non-tribal world ingrained in market economy. The only possible way out in such a situation is, as Kadamanitta understands, the politicization and democratization of such communities. The seemingly significant aspects of science and aesthetics that depoliticizes the environmental discourse is heavily politicized (and in this process challenged) by linking the environmental crisis with socio-political and cultural aspects. Landscape thus acquires significance as a source of livelihood and economic activities as distinct from the mere status as an object of aesthetic appreciation and contemplation. It is in such insights where the focus is not directed to the lost charms, but to the unjust and unequal socio-political and economic organization that the eco-Marxist quality in Kadamanitta's poetry surfaces. Even the sense of desolation that the unprivileged shows for the loss of nature and culture transcends the normal nostalgic/romantic yearning and reaches the radical realization that the downtrodden are never recognized and comprehended. The words of the protagonist of "Kattaalan" reflect the recognition by the marginalized that how the "mainstream" society and cultures have systematically removed the Adivasis from the fruits of nature. Similarly, the desperation and wrath of the marginalized in the regional disparities and exploitation too are subtly expressed:

> When the flower bloomed
> I just got the withering leaves.
> When the fruit reaped
> I just got the skin and husk. (29-32)

A significant aspect of Kadamanitta's poetry is his characters' attitude towards conservation ethics. What is observable here is the special manner in which Kadamanitta's characters relate themselves to their surroundings and to the environmental crisis. Whereas modern conservation ethics is usually projected as being guided by the conflation of ecological sciences and cultural constructs (Trudgill 677), the disbelief, anguish, disappointment and the subsequent wrath of Kadamanitta's people over the loss of their habitat and livelihood, though definitely personal and emotional, is not a bit less genuine. Their particular understanding of their environment is/could be of immense help for conservation activists in ensuring the success of their endeavours. The point here is that activists and ecologists should acknowledge and recognise the specific modes of people's relation to and their sense of and emotions towards nature, especially their immediate environment. What is of importance here is the sentimental and emotional motivation behind people's urge to conserve and sustain their surrounding. Though there is nothing new for us in this notion of conservation ethics, it was not so with 1970s and 1980s, an era in the environmental history of India when creation of wildlife sanctuaries and national parks, even at the cost of alienating people was considered the most effective way to protect endangered flora and fauna.

The protagonists of "Kiratavrutham," "Kattaalan" and "Kurathi" come out of the traditional apolitical structure and question and challenge the outside authority and bare the precariousness of the existence of their communities. It is this aspect of the Kattaalan that inspired T.P. Sukumaran, one of the first critics in the language to theorise on eco-aesthetics in *Paristhithi Soundaryasastrattinoru Mukhavura* (A Preface to Eco-Aesthetics [1992]), to see in him a saviour the Kerala society was looking for. Kadamanitta's Kattaalan, according to Sukumaran, is not an isolated figure for the numerous college dramas staged during the 70s also told the tale of Kattaalan, the Savage (*Nallavanaaya* 55). Sukumaran believes that the Kattaalan symbolizes the urge for protecting nature, Sairandhri, from the intrusion of the civilization, Keechaka

(*Nallavanaaya* 64). This apparent identification of the Kattaalan, Hunter/Savage to the saviour, though appear primitivistic, is far from exalting 'Noble Savage', a concept that cherishes the memories of a pristine, savage and pre-modern existence.

The characters of these poems are essentially revolutionaries in their spirit and aspirations. The words of Kattaalan in "Kiratavrutham," "The hands of huntsman,/I'll chop them with my axe" and the Kurathi's warning: "We'll rise like the fort raised on bones/We'll wake and face you like a castle of stone", forcefully suggest this revolutionary aspect. These utterances of defiance are significant to me as such instances suggest the presence of a universal consciousness of exploitation and the strength to resist it. But, as Sukumaran observes, it is not the responsibility of the literary community to execute the revolution, which is essentially political in nature. The most that the literary community can do is to foster and support the revolutionary attempts (*Nallavanaaya* 61).

Instead of a glorifying, retrospective attitude, these poems yearn for an ecologically and socially just future. Most often this anticipated future is to be attained by means of an abrupt violent struggle. This apparent indifference to the ideals of non-violence is, in the words of Sukumaran, not surprising given the intellectual and material infl uences that characterized the period (*Nallavanaaya* 56). Kadamanitta realized that the injustices prevalent in the then society could not be rectified with the ideals of non-violence advocated by Buddha and Gandhi. Instead he perceived the saviour in the raging Kattaalan.

It would be instructive to juxtapose the defiant and seditious tone of Kadamanitta with the works of Sugathakumari that are evidently educative. Her poem, "Adivasi Saksharatha" (Adivasi Literacy), exposes the ways in which the so called developmental and civilising projects affect the lives of tribal societies and environment. The poem unequivocally states the link between the destruction of the natural environment with the social injustice. The realisation by the Adivasis of the deception of the "benefactors" who destroyed and violated the integrity of their lives and natural environment, unlike the poems of Kadamanitta, does not lead to an encounter, but clearly marks their helplessness. Alongside its reproachful tone, the poem points to the alienation of tribes, animals and other organisms from their natural habitat. The same tone pervades the poem, "Vidhi" (Fate). "Vidhi" presents the Earth as a petitioner who accuses the affluent, powerful elite of the planetary destruction and of the miseries of the wretched and non-human organisms.

These works of Kadamanitta introduced a new aesthetic into Malayalam that shattered notions of propriety by rejecting the thematic centrality of the "mainstream" subject matters in favour of Subaltern concerns. Most of these writings suggest this shift right from their titles. Kadamanitta was following the tradition started by poets like Idasseri. Kadamanitta took the tradition of such writers forward and developed it into a full-fledged aesthetic as well as political concern probing the misery and wretchedness of the deprived. By means of his obvious identification with the Adivasis in these poems, Kadamanitta lends specificity and contemporaneity to the eco-consciousness in Malayalam literature. Besides his apparent identification in these poems with the marginalized, Kadamanitta recognizes that each incursion into the personal as well as into the social life of these groups from the "outside" is detrimental to the environment. The incursions into the natural environment do not go unchallenged, for the protagonists of Kadamanitta are political instead of submissive. Kadamanitta realises that only an active confrontation could counter this transgression. The linking of the natural surroundings to the immediate livelihood of those in the margins, and the realisation that the despoilment of the environment is, as in all instances, detrimental to those who are directly supported by it renders an eco-social colouring to the writings of Kadamanitta. Mostly, groups like lower caste as well as working class women and Adivasis are in direct contact with nature for their livelihood, and by highlighting the concerns of these sections the poems of Kadamanitta proclaim the presence of a powerful subaltern, environmental and feminist aesthetic consciousness in the literature of Malayalam. Kadamanitta, as cited by Coupe, knows well that "it is impossible to separate defence of people from defence of the planet, human rights from ecological survival" and "justice from sustainability" (Coupe 5).

Kadamanitta thus quite forcefully reminds us of the relevance of developing a human rights perspective towards such endeavours, a perspective that eventually became causal to the formation and sustenance of conservation movements like the Chipko that preceded the anti-SVHP campaign and also subsequent campaigns like the NBA. Evidently, this was necessitated by the obvious, though not deliberate, wilderness overtones of the anti-SVHP campaign. What I suggest here is along with the apparent wilderness thrust that the resistance campaign against the SVHP had, it unlike most other wilderness movements was not indifferent to and as is evident from our discussion was even conscious of and acknowledged human rights issues related to environmental movements. The seeming indifference to human aspects, as could be inferred was entirely due to the mere absence of any human settlements in the project area, unlike oppositions to the creation of national parks and sanctuaries that required massive displacement of population or seriously affected their livelihood.[9] Viewed from this perspective, Kadamanitta's nature is not a void, culturally insignificant space where the modern, urban, industrial and capitalistic societies could enact their economic/industrial games. What is suggested here is that nature, even in its so called pristine state, is not free from human interests. Throughout the Silent Valley movement, forest was often cited as an abstract category without any geographical specificities or regional moorings. Forests, in particular, were extolled for their wildness, biological richness and diversity and for their other ecological functions. These perceptions of nature—as an ecologically significant biome or as an economically valuable resource—could be extremely detrimental, as such views hold the potential to turn the public away from forest. What is usually ignored when such perceptions are evoked is the fact that forests, besides being resource or biome, are culturally significant.

Through such reflections, poems of Kadamanitta mount a major offensive against existing socio-economic and political patterns. The plight of their surroundings, to those who people them, is not separate from their own predicament. This understanding of the environmental crisis inspires them to challenge "development," which is often understood as economic growth and technological improvement or specifically as the indiscriminate replacement of forests with roads, houses, business establishments and factories. The knowledge that this transformation of physical geography is closely followed by even more alarming transformations in the lifestyles of those who inhabit such geographies informs and enriches their literary texts. A major discernable transformation is the decline of ecologically viable and sustainable agrarian practises and the ascendance of market-based economies. Hence, these literary texts urge the existing social as well as political systems to seriously rethink and redefine human progress and development in a way that combines reverence of nature with respect for human needs.

Notes and References

1. The term refers to the stream of environmental thought that believes in the existence of an environmentally benign, golden past from which the humanity has fallen and blames the modern, industrial cultures for the contemporary ecological crisis. According to this thought, humanity can redeem nature only by going back to the past.
2. Deep ecology is a strand of radical ecological thought proposed by the Norwegian philosopher Arne Naess that emphasises the intrinsic value of nature as against its instrumental value.
3. For a critical appraisal of these environmental ideas see *Radical Ecology: The Search for a Livable World* by Carolyn Merchant.
4. For a detailed exposition and critique of this position, refer to William Cronon 7-28.
5. Ramakrishna Paniker (1935-), popularly known as Kadamanitta Ramakrishnan, was the most revolutionary among the vibrant group of poets who gathered together for protecting the Silent Valley. In addition to his literary career, he was active in politics too. He was a Member of the Legislative Assembly of Kerala during 1996-2001. Hence, he is sometimes referred to as the Poet-MLA. Till his death in the March of 2008, he was the president of the Progressive Artists Association. He had also served the state as the president of the State Library Council. He was enormously attracted by the ideologies of the Naxalite movement that shook Kerala

during the sixties and seventies of the last century. Marxist ideologies too made a great impression on him. Modern poetry in Malayalam had one of its most important and popular practitioners in Kadamanitta. His poems are lively with the rhythms of folk art forms of Kerala like Padayani. "Kurathi", "Shantha" and "Makanodu" (To the Son) are some of his major compositions.

6. It is exciting to know that the Silent Valley in the Western Ghats is the name sake of a Mountain Park in Northern Ireland situated in the high Mournes that also features a reservoir in the U-shaped valley of the Kilkeel river. As well as being one of Northern Ireland's prime water sources its natural beauty and leisure facilities make it a leading tourist attraction.
7. Though writers like Arundhati Roy has expressed their resentment with the term "writer-activist" for the apparent dichotomy that it proposes between writer and activist, I nevertheless use it to separate writers who were a part of the anti-SVHP movement from its other activists and also from other Malayali writers who stayed away from it. My use of the term is merely descriptive.
8. An indigenous variety of rise.
9. National parks and sanctuaries along with industries have displaced over a million tribes people in India (Ghatak XXVI). Gadgil and Guha describe the way in which creation of sanctuaries interfere with and disrupts the life of local communities (232-239). It could also be recalled here that one of the most publicised and criticised event in the recent Kerala was the forceful occupation of the Muthanga wildlife sanctuary by a group of tribes and the police action to evict them from the sanctuary which resulted in a bloody confrontation.

Works Cited

Bookchin, Murray. *The Ecology of Freedom: The Emergence and Dissolution of Hierarchy.* Palo Alto: Cheshire Books, 1982.

Coupe, Laurence, ed. *The Green Studies Reader: From Romanticism to Ecocriticism.* London: Routledge, 2004.

Cronon, William. "The Trouble with Wilderness: Or, Getting Back to the Wrong Nature." *Environmental History* 1 (1996): 7-28.

Darier, Eric. Rev. of *Natural Causes: Essays in Ecological Marxism*, by James O'Connor. *American Political Science Review* 92 (1998): 935-936.

Gadgil, Madhav and Ramachandra Guha. *This Fissured Land: An Ecological History of India.* Delhi: Oxford UP, 1993.

Ghatak, Maitreya, ed. *Dust on the Road: The Activist Writings of Mahasweta Devi.* Calcutta: Seagull, 2000.

Jagtenberg, Tom and David McKie. *Eco-impacts: The Greening of Postmodernity.* New Delhi: Sage, 1997.

Jayaram, K. "A Photonaturalist in Silent Valley." Manoharan, et al. 51-66.

Kerala Sahitya Akademi. *Vanaparvam*. Thrissure: Kerala Sahitya Akademi, 1996.

KSSP. *"The Silent Valley Hydroelectric Project: A Techno-Economic and Socio-Political Assessment," & "Ecological Aspects of the Silent Valley: Report of the Joint Committee Set up by the Govt. of India and Govt. of Kerala." (combined edition)*. Kochi: KSSP, 1997.

Manoharan, T. M., et al, eds. *The Silent Valley: Whispers of Reason*. Trivandrum : Kerala Forest Department, 1999.

Ramakrishnan, Kadamanitta. "Kattaalan." *Kavitha* Kottayam: D. C. Books, 1984. 31-36.

____. "Kiratavrutham." Kerala Sahitya Akademi 90-93.

____. "Kunhe Mulappal Kudikkarut." Kerala Sahitya Akademi 39-41.

Roy, Arundhati. *The Algebra of Infinite Justice*. New Delhi: Penguin Books, 2002.

Save Silent Valley Society (SSVS). "News Letter II." Thiruvananthapuram. 1-4.

Sugathakumari. "Adivasi Saksharatha." *Devadaasi* Kottayam: D. C. Books, 1998. 26-31.

____. "Silent Valley: A Case Study." Manoharan, et al. 11-20.

____. "Vidhi." *Devadaasi* Kottayam: D. C. Books, 1998. 99-101.

Sukumaran, T. P. *Nallavanaya Kaattaalan*. Kozhikode: Poorna Publications, 1986.

____. *Paristhithi Soundaryasastrattinoru Mukhavura*. Kozhikode: Bhodhi, 1992.

Swaminathan, M. S. "Silent Valley National Park: A Biological Paradise." Manoharan, et al. 1-10.

Trudgill, Stephen. "Psychobiogeography: Meanings of Nature and Motivations for a Democratized Conservation Ethic." *Journal of Biogeography* 28 (2001): 677-698.

Veak, Tyler. Rev. of *Feminism & Ecology*, by Mary Mellor. *Science, Technology, and Human Values* 24 (1999): 402-403.

9

Poppo Pingel and Hugo Kükelhaus—Exploring the Physical Roots of Architecture

—Elmar Schenkel

I

Is there, apart from alliteration, a connection between architecture and archery and archaeology? A rhetorical question indeed, if we look at Poppo Pingel who has been 'arching' in one way or another for the last four decades, but it remains a question to the extent that everybody's life is a questionmark. I have a feeling that archery is the middle term, i.e. that it connects the first and last of the three. The arrow that is shot carries a message from here to there or from there to here, the bow and the arrow actually could be the message. The arrow is often used as an image of time, time that keeps moving on a straight line from beginning to end, or even time that does not move but remains immobile, as in Zenon's paradox. The archer shoots his arrow from the past into the future. He tries to overcome space through speed, but he also tries to overcome time. This becomes especially clear in hunting. The arrow's timing is crucial when the quarry is a moving target. Could one also say that there are shots from the future into the present? The future, like an arrow, is already on the move, it is moving towards us, like the light from a distant nova. It exploded a long time ago, but its light will reach us in the future.

In what sense then are architecture and archaeology connected through time? The architect, at the outset, has nothing but his (or her) imagination. First there is an idea, a mental sketch, an image, and this scant element has to be fully fleshed out by the imagination in an act of construction. An architect, in other words, starts from ruins. Whenever an archaeologist is confronted with a ruin or any other fragment, he or she will have to use their imagination to see the life and the world from which this fragment has come to us. We also call this type of activity reconstruction. Both construction and reconstruction have to complement reality in order to reach another time level of reality: the future or the past. A lot of architectural imagination is based on what architects remember or see in monuments of the past. In poetry, words have to be used that were in use a thousand years ago. The past inspires the future building in the same way as revolutions are fed by ancient models, e.g. Rome. But the future also inspires the present. Our ideas of the shape of things to come influences the way we shape things in the present. The present in its turn inspires our ideas of the past. We begin to highlight the black plague when we ourselves are afraid of epidemics. The way we build or plan to build influence the way we see the lost buildings of Persepolis or Babylon.

II

Architecture requires philosophical inspiration, otherwise it becomes flat and stale and those who inhabit will have to suffer. One of Poppo's inspirations has been the German thinker Hugo Kükelhaus, whose work is hard to define. Was he an architect, a philosopher, a biologist, a carpenter, a designer, a painter or an ecologist? He was all of that and more, one of the last flickers of the uomo universale, the polymath of Renaissance days, an afterthought of Goetheanism in the twentieth century. Goethe certainly was his model in so far as he connected art and science, observation and practice. He designed schools and swimming pools for children and handicapped people, collaborated with architects and developed toys for babies. He designed church windows and furniture, painted walls and created herbal gardens. He was on a permanent lecture circuit on which he inspired thousands and encouraged them to use their own sensory experience. He wrote books on inhuman architecture and on the role of the body for buildings. He also developed what was to become fields of sensory experience (*Erfahrungsfelder für die Sinne*) which are now to be found in places all over Germany and Switzerland. In these fields one can learn about one's own senses and their importance for human and even moral development. It is important to note that they are not simply stations at which one can study the laws of nature and thus understand scientific facts. They are rather meant to reconnect ourselves to our sensory experience, an experience that has been blocked by technology, education, architecture and other factors.

Hugo Kükelhaus (1900-1984), though travelling extensively, lived for the last decades of his life in the medieval Westphalian town of Soest, between Dortmund, Münster and Paderborn. My father, a craftsman, who had introduced archery, in particular Japanese archery to our village near Soest, knew Kükelhaus and it was through him that his nephew Poppo and myself came to know him. It should be mentioned that my father also had an important influence on Poppo in his early youth as far as attitudes to life, work and philosophy were concerned. And of course Poppo also learnt archery. After my father's death in 1974 I got in touch with Kükelhaus, who employed me as an assistant at his Munich exhibition where he introduced his sensory field to the public (1975). I stayed in contact with him, wrote about him, saw him and organised a lecture for him. When Poppo came on one of his annual trips to Soest he went to see Kükelhaus and his wife. This visit, the only time he saw him, took place in 1982 and impressed him deeply. As Poppo remembers:

> On a cold winter morning of 4th November 1982 I saw Hugo Kükelhaus for the first and last time, high up in the attic of his 300 year old timber frame house in Soest, Germany. He was 82 years then. I had questions regarding life, architecture, sketching and calligraphy at the back of my mind when I met him. However, our discussion or rather his monologue revolved around his resistance against Hitler, Einstein's theory of Relativity, life, cosmos and its curves. Charged by his energy I didn't even turn on the heater all the 60 km driving back home. Those two hours with him and his personality left me deeply moved.[1]

It was a Poppo style visit his relatives and friends have been familiar with: impromptu, no appointment, out of the blue. Mrs

Emilie Kükelhaus let him in und led him up into her husband's famous attic which is still to be seen in the 17th century timber frame house, Nöttenstrasse 28 in Soest. Poppo's visiting card was his book on Rural India, in which he described how to build mud cottages. Of course, it was not a mere pretext. He hoped to get some enlightening commentary from Kükelhaus. Kükelhaus was grateful, but then got carried away with his own philosophy. He taught Poppo his ideas in a nutshell, continually showing things to him, geometric models, archeological finds, stones, drawings, Buddha statues. Poppo listened, enthralled. Kükelhaus, too, was impressed with his visitor from India, as I was to learn afterwards from him. This mixture of Westphalian sturdiness and Indian background was the kind of thing that he enjoyed and admired. Eventually, Kükelhaus gave a few nibs for calligraphy to Poppo. This gift is an important sign, since both are great lovers and practitioners of longhand writing. Like archery, pottery, ikebana, kendo, like cooking, painting or gardening, calligraphy is an art of self-fulfilment. Beautiful writing can be an end in itself—especially important these days when we are forgetting it rapidly and keep pushing buttons instead—but it can also be a way of expressing the harmony between the self and the universe, or of approaching this. At the end of the visit, Kükelhaus drew a famous geometrical shape on a piece of paper many of his followers or readers will know: it shows just the three angles of the triangle without the connecting sides. The effect is powerful: one feels a strange vibration on the paper between the three points. The gaps are replaced by the spectator who tries to connect the shapes. This philosophy of the imperfect and openness is central to Kükelhaus' thought. It is also at the basis of his criticism of the modern world where everything is consumed in a readymade form. A toy, for example, should leave room for the child's imagination, otherwise this will wither. It also applies to our information overload which makes us unhappy, to say the least.

Kükelhaus inscribed this sheet of paper with the following words: "Diese Richtung ist gewiss, immer schreite, schreite, schreite ... Hindernisse und Finsternisse bleiben dir beiseite." Kükelhaus liked to quote Goethe in all kinds of situation. His written works abound with quotations and his whole outlook points to the great classical writer and thinker. In his sensory field, he put up a station at which one can explore Goethe's colour theory. When, on one occasion, journalists asked him how he had "hit upon Goethe", he replied: "You had better ask how Goethe hit upon me!" This was considered an arrogant answer, which in fact it is not. It is the opposite. Goethe, if we follow this line of thought, had covered everything, from poetry to the intermaxillary bone, and thus had also covered his followers, had foreseen Kükelhaus. Be that as it may, the Goethe quotation Kükelhaus gave to Poppo ended up in India. It was translated into Gujarati and now decorates the *chaikhana* of Poppo's and Mona's house in Auroville.

Though it was the only time that Poppo met with Kükelhaus, it was not the first time he had gone to Kükelhaus's place in Soest. In fact, he and my father Heinrich Schenkel went to see Kükelhaus in 1965. Poppo had brought along an Afghan friend of his from Peshawar. It could have been a significant encounter, but, alas, Kükelhaus was out of house. His wife was there, however, and she showed her visitors around. Poppo was deeply fascinated by all those tribal objects, Mexican or African things, crystals, mummified hands, pyramids, inscriptions and especially the Buddha's head. And it was then that he asked Mrs. Kükelhaus the ponderous question: "What does your husband actually do for a living?" ("Was macht Ihr Mann eigentlich?").

Picture *Chaikhana* Skip

What was it that attracted Poppo in Kükelhaus' work and personality? It was a characteristic blending of philosophy and the practical, I think, that he found in Kükelhaus and that he had striven himself for. Like Kükelhaus, Poppo was a craftsman from the start, but one who was looking into and simultaneously through his craft. In both cases it was carpentry. Both were seeing the wider contexts of their craft and also its exemplary nature. No matter what you do, it can become a symbol or a mirror to universal forces if you do it in the right mind. Both were looking for this 'right mind' when working with their hands. Kükelhaus's practicality was always suffused with charisma, he was down to earth and at the same time talked about the galaxies. What is above is also below, as the *Esmerald Table of Hermes Trismegistus* says. Kükelhaus saw the same galaxy (a Greek word for milk) in his cup of coffee. He conversed with sober Westphalian farmers and artisans, but also with philosophers and physicists from abroad. The arrow was in the centre when he spoke. In this, he was like a Daoist.

Yet we must not ignore Poppo's Indian background. He came from a similar world as Kükelhaus when he left Germany for good in the 1960s and whenever he returned he felt at home and enjoyed his roots in the village of Oestinghausen and the family in Wiltrop and Münster. He could relate to that easily, the world of practical people. At the same time India, and in particular Aurobindo and the Mother had a decisive influence which made him see his own roots in a different light. Poppo's Westphalian roots are deeply embedded in the material world, in its earth and stones, in the fields and the rural work, but also the provincial cities and their strict Catholicism. His roots defined him up to a certain extent and it needed the Indian influence to liberate him from them, at least in part. However, his connection to Kükelhaus and Westphalia has remained vital. Kükelhaus remained for him what he called his "material guru" (not *materialist guru*, obviously). Like Goethe, he pointed to the janus-faced gates of the senses: they are embodied and yet transcend the body. This is a reason why so many mental activities are expressed in bodily terms—to grasp, conceive, understand et al.

After Poppo's 1982 visit, Kükelhaus became an important topic in our letters and conversations. It was Poppo who around 2000 instigated me to write a translation of what we thought was Kükelhaus's most important book: *Inhuman Architecture*. In this, he attacks the kind of mass produced buildings of the 1960s and 70s, especially in the educational area. He castigates windowless rooms, air conditioning, artificial lighting and in general an architecture that is enemy to the body and its needs. If this is the environment we are making for the next generation, then woe to us and them! Poppo thought that Kükelhaus's ideas had already entered mainstream thinking in German architecture—at least architects and contractors know more about this than 30 years ago—but that they were especially needed in countries with an enormously accelerated urbanisation and growth rate like India and China. All the mistakes made by the West in the 1960s are being repeated on a much grander scale now and thus it is all the more important to hear the warnings of thinkers like Kükelhaus.

Hence I started to translate and this activity became a reason for two of my trips to Auroville in 2000 and 2004. While Poppo and his people were digging in the archeological excavations at Auroville I was sitting in the shade of a eucalyptus tree and working on the strange transfer of German phrases and words into English. India and Auroville were an excellent place for this job. In the light of this country, Kükelhaus's work gained an extra-dimension, it became more universal. I also enjoyed the possibility of discussing difficult phrases or terms with Mona and Poppo. Every evening, we would withdraw for a while and check my day's work. I was in particular need of their baubiological knowledge in order to clarify would Kükelhaus actually meant in German. We also had to straighten and shorten his lengthy and sometimes unwieldy German sentences to make them more palatable for the English language reader.[2]

III

It was a worthwhile experience, hopefully not only for myself. I was especially intrigued by the coordination and cooperation of words, spades and buildings, in other words by language, archaeology and architecture. Digging for words and constructing a text all went hand in hand. Language here took over what archery did metaphorically: connecting and shaping things in time, even connecting archaeology and architecture. Words are our means of communicating the imagination and there would be no idea of the past or of the future without words. However, for Poppo and for Kükelhaus language only has or had a very limited range and in this sense, their attitude is comparable to the early theologians of whom Alain de Botton wrote in his book on architecture:

> ... the early theologians speculated that it might be easier to become a faithful servant of God by *looking* than by *reading.* They argued that mankind could more effectively be shaped by architecture than by Scripture. [...] We might learn more about humility by gazing at an arrangement of tiles than by studying the Gospels, and more about the nature of kindness in a stained-glass window than in a holy book.[3]

References

1. De Botton, Alain, *The Architecture of Happiness*. London: Penguin 2007,
2. Kükelhaus, Hugo. *Inhuman Architecture*. Studio Nqshbandi: Auroville 2007.
3. Schenkel, Elmar. *Ein Lächeln und zwei Fragezeichen. Indisches Reisetagebuch.*. Eggingen: Edition Isele 2001.

10

Many Ecologies of the Self: Inner and Outer Spaces in the Poetry of Kamala Das/ Suraiya, An Ecofeminist Reading*

—Usha V.T.

The contribution of Kamala Das/Suraiya to the corpus to poetry written in English in India is no minor one. This has been amply borne out by the large number of contemporary critical readings of her work which have looked at various aspects of her poetry and poetics and the equally large number of learned dissertations and critical expositions of her work. However, this attempt to read her poetry from an ecofeminist point of view traverses new grounds and helps provide new insights into hitherto unexplored regions in her poetic world such as the connections between the human and the non-human world of her poetry, her celebration of the body, and male-female dichotomies. This paper attempts to draw attention to the blatant violation of conventional hierarchies through her poetic voices and the placing of her poetic personae within their specific ecological environment which make her poetic output ecofeminist. In addition it attempts to identify her capacity to integrate binary oppositions and speak from within alternate viewpoints. The prime persona in her poems is herself, which she locates either in a physical environment or in relation to other personae.

It may be argued that the poetics of Kamala Das lends itself to ecofeminist reading mainly on account of the obsessive presence of some of the following elements. Though concerned with oppositions, such as nature-culture, male-female, body-soul, she does not treat them in absolute terms or as contradictory or mutually exclusive. She is able to absorb the seeming contradictions into a composite picture. She does not limit them into hierarchical power structures wherein one is superior to the other. She is deeply concerned with the social situation of women, and how their physical bodies have been reduced to being mere commodities or products for general consumption. Her poetry amply reflects her intense internal struggle against the reduction of the female into being a mere body and the patriarchal society's victimisation of women. Women held little space in a social system that was solely the male world! Nevertheless, in her poetry, one can also find an appreciation of the intrinsic value of everything in nature— a respect for the natural, a biocentric rather than anthropocentric approach to the natural world.

Descriptions of nature and the outer world have always been abundant in Indian Poetry. Whether one is looking at the large body of Sanskrit poetry or poetry in any of the regional languages, it is impossible not to see the various moods and cadences of nature reflected therein. It would then be almost unfeasible for English language verse in India to be otherwise. Many of the early Indian writers in English wrote in a somewhat imitative vein—under the influence of the British Romantics—"Mathew Arnold in sari"[2] (as Gordon Bottomley infamously critiqued them!). However, it is significant that they drew inspiration from the flora and fauna around them for their poetic utterances. Among the women writers the voices of Toru Dutt (1856-1877) and Sarojini Naidu (1879- 1950) are laudable instances reflect this. In their poetry, the ubiquitous images of the European Rose or Lily was replaced by the Indian Lotus and the essence of Indian culture began to gradually unfold itself through their verse when they wrote of *koels* (the Indian cuckoo) and Champak blossoms, referred to Krishna, the Dark God, or the snake charmers or the women in purdah. The language of their poetry however continued to be largely repetitive and imitative with less and lesser self reflexive qualities.

It was only with the publication of the poetry of Kamala Das, now Suraiya that Indian poetry took an intensely serious personal tone, a self reflexive language and earned a prominent place among world poetry in English. C.D Narasimhaiah, noted Indian critic and scholar, describes her poetry thus: "She is perhaps the only Indian poet who owes little to Yeats or Eliot and trusted to her own resources and to her culture— thanks to a poet-mother and her indefatigable Keralite upbringing, it is possible she felt re-assured in the opulence lying all around her to kindle her imagination"[3]. The poetry of Kamala Das draws unhesitatingly from her matrilineal heritage, the land, the environment as well as the poetic personae and the personal voice.

Born in 1943 in Malabar, now Kerala, a small state in the South of India, Kamala Das grew up in her matrilineal home and was influenced a great deal by the literary influences around her. Her maternal uncle Narayana Menon was a poet and her mother Balamaniamma a well- known poet in Malayalam. She was married in 1949 and consequently left Kerala and her matrilineal home to accompany her husband who was stationed in Calcutta, New Delhi and Bombay. They returned to Kerala in 1981 on his retirement. Her first collection of poems *Summer in Calcutta* was published in 1965. This was followed by *The Descendants* (1967), *The Old Play House and Other poems* (1975), *Tonight, This Savage Rite: The love poems of Kamala Das and Pritish Nandy* (1979), *The Anamalai Poems* (1985). Her autobiography *My Story* (1975) evoked much criticism particularly in her home state Kerala, where the expressions of physicality and sensuality in her poetic utterances were treated as expressions of her waywardness. She has of late converted to Islam and rechristened herself Kamala Suraiya

The poetic utterances of Kamala Das have a very personal note. The main protagonist of her poems is her self. The poetic voice is confessional. She makes no effort to be objective and scientific although her descriptions are accurate, precise and quite detailed in their content. She is clearly subjective in her approach and makes no effort to rationalize her emotions. Yet she does not really sentimentalise the self. In "Composition", an early poem she writes:

I also know that by confessing
By peeling off my layers
I reach closer to the soul
And
To the bone's supreme indifference.
Only those who like to listen,
Listen.
What I narrate are the ordinary
events of an
ordinary life.

Yet as poet she transforms the commonplace of the ordinary into something splendorous. The ordinary events of her personal life comprise the corpus of her poetic world. She is able to transform the personal into the political in a very unique manner using a very subjective tone. In her poetry the private world of her self is politicized giving uninhibited expression to "the full range of female experience"[4]

Dress in sarees, be girl
Be wife they said. Be embroiderer, be cook,
Be a quarreller with servants.... Don't sit
On walls or peep in through our lace-draped windows
Be Amy or be Kamala. Or better still be Madhavikutty.
It is time to
Choose a name, a role. Don't play pretend games.
("*An Introduction*")

Though the self of her poems is asked to "fit in" she finds that she cannot fold herself into the neat compartments she is asked to, for she is more than just Amy or Kamala or Madhavikutty (her penname in Malayalam is Madhavikutty, the wife of Madhava das,). She finds that she is more than all that for she is "every woman who seeks love". She is direct and outspoken in her poems. Her poetic personae are located in the India of the present with all its colonial complexities. In "An Introduction" she introduces her poetic persona thus:

I am Indian, very brown, born in
Malabar, I speak three languages, write in
Two, dream in one.

She is quite aware of the inherent contradictions therein, yet she is able to reconcile these contradictions within her poetic frame for she goes on to state her blatant distrust of accepted definitions and advice. She prefers to keep to her own decisions and explains her poetic stance in the poem when she demands boldly:

> Why not let me speak in
> Any language I like? The language I speak
> Becomes mine, its distortions, its queernesses
> All mine, mine alone. It is half English, half
> Indian, funny perhaps but it is honest,
> It is human as I am human, don't
> You see? It voices my joys, my longings, my
> Hopes, and it is as useful to me as cawing
> Is human speech, the speech of the mind that is
> Here and not there, a mind that sees and hears and
> Is aware. Not the deaf, blind speech
> Of trees in storm or the monsoon clouds or of the rain or
> the
> Incoherent mutterings of the blazing
> Funeral pyre.

In these lines one can see the directness of the tone of her poetry, its refusal to conform to traditionally accepted metrical patterns and its subjective themes. But despite its preoccupation with the self, it deals with the outer world and the natural phenomena with respect and a certain sense of empathy. The human world is co-terminus with the non human schemata. The birds, the animals, the trees, the storm and the rain are as much a part of her inner world as her emotions and opinions.

In many of her poems, Kamala Das reverts to the world of her daily life. Stray incidents from her childhood, her school days, her teenage fantasies, her marital and extramarital exploits—imaginary or otherwise, her family life, her numerous relatives but mainly her grandmother are some of the major themes and personnae of her poetry. Her grandmother and the matrilineal home in which she lived are living personae in her poems. They assume a living quality, mixing memory with desire" to relive a past now lost to her. In a short but powerful and much anthologized poem "*My Grandmother"s House*", she describes the house and her attachment towards it thus:

> There is a house now far away where once
> I received love.....That woman died,
> The house withdrew into silence, snakes moved
> Among books I was then too young
> To read, and my blood turned cold like the moon
> How often I think of going
> There, to peer through the blind eyes of windows or
> Just listen to the frozen air,
> Or in wild despair, pick an armful of
> Darkness to bring it here to lie
> Behind my bedroom door like a brooding
> Dog........you cannot believe, darling,
> Can you, that I lived in such a house and
> Was proud, and loved ......I who have lost
> My way and beg now at strangers doors to
> Receive love, at least in small change?
>
> —(*The Old play House and other poems,* 1973, 2004) p.32

The house becomes a part of her past and the qualities of love and affection she associates with her grandmother make it a living entity with a capacity to love that she cannot find in her urban present. Here the persona of her self with a personal history becomes unified with her poetic subject. The other persona of the poem, the snakes, the books, the moon and the windows of the house, as well as the darkness and the Dog become unified as symbols of her loss. There is no significant distinction between the human and the non-human. The inanimate windows and the darkness "like a brooding dog" assume living qualities, like the house which makes her proud of the fact that she was loved. Her matrilineal home, from which, she was relocated by marriage, becomes the essence of the emotions nostalgically associated with her childhood and youth and she sees herself as a lost wanderer disinherited from her matrilineal past, begging at stranger's doors now for her lost inheritance of love. These poems could also be treated as a socio-cultural record of women within matriliny in Kerala.[5]

The landscape of the land wherein the poems are located become an inevitable and integral part of her poetry. Her matrilineal home (the *tharavad* also called the *nalukettu*) appears and reappears in several poems. An interesting feature with regard to the tharavad, is the life style practices wherein the natural environment extends into the house. There is a pond attached to every house as well as a large natural garden wherein a space is allotted to the worship of the elements and the non-human creatures like the snake are allowed a safe haven from human disturbances. They are not only left unharmed but are even venerated and there are specific areas in the compound of every house where shrubs and trees are allowed to grow wild and the natural habitat is left undisturbed. The house itself is an architectural marvel that allows for natural ventilation with large spaces within, open to the sky and low sloping roofs that are eco-friendly, and suitable to the local climatic conditions. It is therefore only inevitable that poet extends those eco-friendly approaches to her poetic subject matter. She speaks of her love of swimming, which "comes naturally to me" and the opportunities she had in the tharavad to indulge her passion:

> I had a house in Malabar
> I did all my growing there
> In the bright summer months.
> I swam about and floated,
> And dived into the cold and green
> I lay speckled green and gold
> In all the house of the sun
> Until
> My grandmother cried
> Darling you must stop this bathing now
> You are much too big to play
> *Naked in the pond.*
>
> (*"The Suicide"*)

The sensual pleasure of the swimming in the green pool with all its visual spectacle ends abruptly, when she is made aware of the growth of her body and her transformation into a young woman. The structure of the poem with its uneven rhythms broken at that juncture reflects the break in her sensibility as well. The sudden awareness of her growing body, which is a culturally imposed one and its implications are reflected through a discontinuity from the natural world around her.

Love is a major preoccupation in the poetry of Kamala Das. She mourns the loss of innocence and the forfeiting of relationships based on love in several of her poems. Yet it is the physical and sexual expression of love in her poems that has attracted much critical attention. As Bruce King remarks, her poetry has "created a climate for a more honest revelatory, confessional poetry by Indian women..."[6] Her candid descriptions of the physicality of the human body in many poems have distressed many critics. Her blatant descriptions of the human body and the inherent sexuality of male- female relationships have brought her much unwarranted criticism.

> Gift him all,
> Gift him what makes you woman, the scent of
> Long hair, the must of sweat between the breasts,
> The warm shock of menstrual blood, and all your
> Endless female hungers.
>
> *("The Looking Glass" The Descendants)*

The importance of celebrating the body is highlighted by ecofeminist writers like Carolyn Merchant when she obliterates the very difference between the self and the environment by referring to the body as the "primary environment". "Reclaiming the body is important work for eco-feminism, particularly since human (male) identity has been equated with the mind alone and the body has been variously raced animalized, feminized and naturalized in order to be seen as inferior and antagonistic to the progress of culture"[7]. In the poetry of Kamala Das both the male and the female body are explored and celebrated. Her poetic voice reflects ecofeminist qualities when in the same poem for instance the female persona advises the reader

> Admit your
> Admiration, Notice the perfection

Of his limbs, his eyes reddening under
Shower, the shy walk across the bathroom floor,
Dropping towels, and the jerky way he
Urinates....

But her approach to male- female relation is not merely physical or dualistic. The mind or the soul is as important as the body as she makes clear in poem like "Suicide" where she states:

Bereft of soul
My body shall be bare.
Bereft of body
My soul shall be bare

Both her body and her mind are the subject matter of her poetic utterances. She is explicit and unashamed about her inner longings and her physical and "sexual hungers" in her poems inviting much adverse critical comment from conservative Indian critics, who decry her "endless female hungers". The persona of her poetry would then need a body as well as the spirit and the environment that supports the whole to reflect her holistic vision. At other instances, she would rather look at the spirituality of the emotions by sublimating them as in the Radha-Krishna poems. In the poem "Substitute" for instance, the female narrator says:

Yes I was thinking, lying beside him
That I was loved, and was much loved.
It is a physical thing, he said suddenly,
End it, I cried, end it, and let us be free
(*Only the Soul knows how to sing*)[8]

Love therefore was not to be seen in purely physical terms, yet there was to be no denial of the body or its sensuality in her poetic voice. Sudesh Mishra, for instance, links the quality of love in her poetic utterances to the bhakti tradition by saying that " the Radha krishna cult of love making, a genre of its own right within the tradition of bhakti verse, affords a unifying dialectic to her poetry."[9] So whether her idealized or Divine lover continues as Krishna or is transformed into "Ya Allah", the spirit with which she addresses Him remains the same and the reader is transported to a uniquely feminine world. In one of her recent Laurentian poems she writes:

> Ya Allah
> I thank you once again
> And again and again
> For this gift of a man
> Who is now my sustenence
> The draught I thirsted for
> And the sole raiment
> For my nudity, both my body's
> And my soul's.[10] (" The maples are Green Still")

At the same time, many feminist critics are also suspicious of the authenticity of her "female" voice. Herein lies her unique position vis a vis the female voice in modern Indian Poetry in English.

She has been often been referred to as a feminist though she herself is unsure of the appropriateness of the label. In a recent essay, she states" When people hint that I am comely despite my age, I blush in gratitude and discover within myself that I am no feminist writer entitled to inaugurate a women's meet".[11] Yet the personae of her poetry reflect a very overt feminist stance. The ouvre of her poetry is undoubtedly a woman's world.

When Virginia Woolf wrote distinguishing the public and the private worlds she drew attention to the dual world of a woman in the modern societal perception as well as the values attributed to each of these.

> The world is at present divided into two services: one the public and the other the private .... Is the work of a mother, of a wife, of a daughter worth nothing to the nation in solid cash/ ... astonishing[12]

Kamala Das has no hesitation in placing herself among the world of the home with its domestic rights and duties. And she is far from apologetic in her stance when she does so. For her the world of her home is in continuity with her public self. She does not look at them as binary oppositions.

The environment around the personae of her poetry is very much a part of their selves or herself. At times the outer space is in sympathetic union with the inner self of the personae. At other times there is no marked empathy. She does not idealize the land or its many manifestations nor does she alienate herself from it. The environment of the poetry is never in direct confrontation with the speaker/s.

> Ghanshyam,
> You have like a koel built your
> Nest in the arbour of my
> Heart
> My life until now a sleeping jungle
> Is at last astir with
> Music
> (*Tonight this Savage Rite* p.18)

The use of the name Ghanshyam, one of the numerous pseudonyms for the Lord Krishna,

Whom she addresses as her Divine lover, is significant. The name may be translated as the dark one heavy like the dusk or the dark cloud that heralds rain. The lover is equated to a phenomenon of nature the dark cloud heralding rain, the raindrops of course symbolising fertility. Herein the self and the other, the lover and his love become one with the natural phenomena around the poetic persona.

In the final analysis Kamala Das's poetry as I have argued is seeped in ecofeminist concerns. Kamala is of course not a "nature poet". She writes of herself however, not in the singular but in the plural. Kamala becomes every woman. Kamala's self becomes, the ecological self. Her poetry has celebrated the body with such zest that a critic like M.K.Naik states that "the total impression Kamala Das's poetry provides is one of a bold ruthless honesty tearing passionately at conventional attitudes to reveal the quintessential woman within."[13] Through her poetic voice she was able to celebrate experience of being woman without apology or shame. Through her poetic lens, she looks at various oppositions like the male-female, nature-culture and human-non-human and is able to arrive at an integrated and composite picture rather than a broken or discontinuous one. She does not gloss over differences or ignore them, but tries to look at them with an empathy that is non-dualistic and holistic. While Kamala Das looks at these dichotomies, she is also highly conscious of the dependency existent in heterosexual relations, their interdependent and hierarchical relationship as well as their need for individuality and separateness. In structure, theme and ouvre, the poetry of Kamala Das/ Suraiya lends itself to ecofeminist readings.

Notes and References

1. Quoted in K.R Srinivasa Iyengar *Indian Writing in English* (Bombay: Asia Publishing House, 1962 p7)
2. *Introduction to An Anthology of Commonwealth Poetry. Ed. C.D.Narasimhaiah.* (Madras: Macmillian, 1990) p11
3. Prefatory note *Perspectives on Kamala Das's Poetry* ed. Iqbal Kaur. (New Delhi: Intellectual Publishing House, 1995)
4. Ref Robin Jeffrey, *Politics, Women and Well Being* (1993), wherein the historian examines the sense of well being experienced by the women in Kerala in their matrilineal home and the consequent benefit to the community at large and my own paper on "Matriliny and the Kerala Woman" *South Indian Studies* ed. S. Murali (Delhi: B.R Publications, 1998) pp 51-58.
5. Bruce King *Modern Indian Poetry in English.* (Delhi: OUP,1987) p155.
6. Greta Gaard Introduction to Special Issue on Ecofeminist Literary Criticism *ISLE* 3,1, (Summer 1996).
7. Kamala Das *Only the Soul knows how to Sing* (Kottayam: DC Books, 1996) p97.
8. Sudesh Mishra. *Preparing Faces: Modernism and Indian poetry in English*, Univ of South pacific and CRNLE, 1995 p 353.
9. Kamala Das/Suraiya "Laurentian Poems: Quebec, Canada" in *Kavyabharati* 12,2000. Special Issue Poetry of Indian Women.
10. " Body Awareness is by itself a weakness. Gender Awareness is Worse" in Life/Style Rediff on the net. http://www.rediff.com/style/sep/16das.htm 25 April 2008.
11. Virginia Woolf, *three Guineas,* 1992, 230.
12. M.K Naik A History of Indian English Literature New Delhi, Sahita Akademi, 1982.p.210.

* Parts of this paper were presented at the International Seminar on Ecopoetics,University de Libre, Brussels 14-17 May 2008

11

Body as Colony: An Ecofeminist Analysis of Margaret Atwood's *The Handmaid's Tale*

— K. Reshmi

> We are for breeding purposes [...]
> We are two-legged wombs,
> that's all: sacred vessels, ambulatory chalices. [HT 128]

Ecofeminism, which combines ecology and feminism is a concept which questions the patriarchal oppression of women and the capitalistic exploitation of nature. As Greta Gaard states, ecofeminism's basic premise is that the "ideology which authorizes oppressions such as those based on race, class, gender, sexuality, physical abilities is the same ideology which sanctions the oppression of nature" (Living Interconnections" 1).

In *The Handmaid's Tale*, Atwood deals with various ecofeminist issues in a brilliant and effective manner. Science, with its reductionist policy, has indiscriminately exploited nature for man's benefit. Deforestation, industrialization, hazardous nuclear testings and environmental pollution have far reaching consequences on the health of the people and the biosphere. In *The Handmaid's Tale*, Atwood points out how such a situation can affect women more than men. Atwood discusses how female body can become a territory to be colonised by the patriarchy, how female sexuality and reproduction can come to be regulated by men. In such a system, woman, like nature, becomes a national resource. The woman's colonised existence is related to the region's—Canada's—similar situation. Atwood employs ecofeminist metaphors to emphasise the theme of oppression of woman and nature and makes the colonised female body the body of the narrative.

The novel is set in 2195 in Cambridge, Massachusetts, formerly known as the Northern United States of America. It dramatises a futuristic, bleak and totalitarian society where women are denied the fundamental rights. The Republic of Gilead, the futuristic society that Atwood creates, is governed by a fundamentalist Christian theocracy. They have shot the President, machine-gunned the Congress and the army has declared a state of emergency. The Constitution is suspended, universities are closed, professors or lawyers are not there any more. In *The Handmaid's Tale*, Atwood depicts a horrible regime, which is both racist and sexist in its views.

The reasons for the formation of Gilead can be traced back to the ecological degeneration of the present times. As Professor Pieixoto remarks in the Historical Notes of *The Handmaid's Tale,* there are several reasons for the decline in population:

> Some of the failure to reproduce can undoubtedly be traced to the widespread availability of birth control of various kinds, including abortion, in the immediate pre-Gilead period.[...] Still births, miscarriages and genetic deformities were widespread and on the increase and this trend has been linked to various nuclear plant accidents, shut down and incidents of sabotage that characterized the period, as well as to leakages from chemical and biological-warfare stockpiles and toxic-waste disposal sites [...] and to the uncontrolled use of chemical insecticides, herbicides and other sprays. [HT 286]

Due to the indiscriminate use of reproduction controlling technologies, due to sexually transmitted diseases and environmental pollution of the present times, there is an enormous decrease of population in the society of Gilead.

The consequences of this are suffered by the women of Gilead. In this totalitarian regime, women are denied the right for citizenship. A new law has been enforced that women cannot hold property; it should be transferred to the next male kin—husband or another one. Women are not allowed to go for employment and their bank accounts are freezed. Women with viable ovaries are taken as handmaids by the otherwise childless Commanders to augment their species. Each handmaid is given three chances, each of two years with one Commander. If a handmaid fails to produce a child within these six years, she will be declared as Unwoman and exiled to the colonies where she has to clean toxic wastes. If the handmaid successfully completes her term by producing a child, her child will be taken up by the Commander and his wife and she is then transfered to the next Commander. Thus, in the Gileadean regime, woman is treated as a national resource.

Ecofeminists question the patriarchal attitude to objectify woman as the 'other' of man. According to them, this sort of dualistic thinking results in hierarchy and oppression. As Simon De Beauvoir points out, man defines himself as the 'self' and woman as the 'other'. He is the subject, the mind, the centre of reason, active form and she is the object, the body, the centre of emotions, mere passive matter. Not only woman, but also nature, has been identified as the 'other' of man. Zuleyma Tang Halpin argues in "Scientific Objectivity and the Concept of 'the other':

> Women have been oppressed not so much because they have been equated to nature, but rather because both women and nature have been equated to "the other" [...] the same dynamic that has resulted in labelling woman as inferior and justified society's domination of women and nature, has done the same during most of our history, to Blacks and other people of colour, Jews, the poor and gay or lesbian persons. (qtd. in O'Loughlin 148)

It is significant to notice that nature has also been imaged as female. Woman is equated to nature because of her life-giving activities. Ecofeminists are of the view that the claim for the superiority of 'self' over 'other' is based on the difference between the two. This difference is based on the assumption that the self possesses things which the other lacks. This alleged superiority of the self, moreover, is used to justify the subordination of the other. In the patriarchal society of Gilead, woman is reduced to the status of an object and is denied the power of reason, emotion, thought and individuality.

The handmaids in Gilead do not possess even a name. Absence of a real name signifies a lack of identity. They are named after the Commanders with whom they have to perform the impregnation ceremony; thus their names are Offred, Ofglen, Ofwarren, Ofwayne and so on. These names indicate that the handmaids are designed for a particular function and are defined only in relation to a man. In fact, Offred's real name is never mentioned in the novel. She knows that the name, like her existence, is useful only for somebody else: "Your name is like your telephone number, useful only to others" (HT 80). To quote Jessie Givner, "Indeed, the desire of the Gilead regime to remove names is as strong as the desire to remove faces. Just as the rulers of Gilead try to eliminate mirrors, reflections of faces, so they attempt to erase names" ("Names, Faces" 58).

Women in Gilead are colour coded based on their relationships to men. Wives of the Commanders are dressed in blue, Daughters in white, Marthas in Green and Aunts in Khaki. Econowives, the wives of poor men wear red, blue and green striped skimpy robes. The handmaids are uniformly dressed in scarlet robes which signify adultery. Their dress is long and concealing with white wings and red veil. Through this dress code, the Gileadean regime pushes down handmaids to anonymity and invisibility, hence Aunt Lydia says, "Modesty is invisibility" (HT 43).

The patriarchal set up of Gilead denies women the power of word. They are prohibited from reading and writing, the penalty for trying to do so is severe. For such acts of disobedience, their limbs are chopped off mercilessly as those parts are not necessary for the purpose for which they are reared. The handmaids communicate mainly through gestures and are permitted to talk only if it is most necessary. Even the shops are identified by pictographs and not names. Thus the handmaids are relegated to a languageless realm in order to deny them power. The ruling class knows that words are weapons which can free people from bondage and subjugation. In her lecture "An End to Audience", included in *Second Words*, Atwood articulates this idea:

> In any totalitarian takeover whether from the left or the right, writers, singers and journalists are the first to be suppressed ... The aim of all such oppression is to silence the voice, abolish the word, so that the only voices and words are those of the ones in power. Elsewhere, the word itself is thought to have power, that's why so much trouble is taken to silence it (350).

The handmaids are completely objectified and are denied the power to judge and reason. They have no freedom to decide their own needs and activities. Their bath is regulated by others, their food is chosen by the authorities based on its nutritional value. They have no right to refuse food even if they are not hungry. Thus, they are like animals reared just for breeding or commercial purposes.

The handmaids are totally separated from their roots and are not permitted to have any contact with their past. So Offred says, "I am a refugee from the past" (HT 213). Offred was university-educated; she loved books and worked transferring books to computer disks. She also remembers the time she spent with her husband Luke and her daughter, who are now missing. Offred is totally uprooted from her past and the only reality before her is the present—that she is thirty three years old, has brown hair and stands 5 feet 7inch without shoes. Not only handmaids, but also the Aunts, Marthas and Wives are likewise separated from their past. Nobody mentions the past as memories can only bring pain. So, Offred says that in Gilead, "thought is rationed" (HT 7).

The handmaids are denied all emotions and are expected to passively perform the functions ascribed to them. All sorts of entertainments are prohibited in Gilead. They are allowed to watch movies once a week, after lunch, before nap, but these movies are for educative purpose. These are the old pornographic films of the nineteen seventies and eighties depicting the brutalities acted on women. The purpose of showing such films is to create terror in their minds regarding the pre-Gilead period and to assure them a sense of false security in their present mode of living.

In the totalitarian regime of Gilead, sensuality is repressed and the word love is prohibited. The desires of the individuals are restricted by state regulations. The Aunts constantly warn the handmaids that all flesh is weak and it is the duty of women to set limits. In Gilead, sex has been reduced to mere reproductive duty without love. This puritanical sort of environment makes her crave for sensual experience and hence, her heart easily turns to Nick.

The handmaids, having been denied thoughts, emotions, desires, are denied individuality and are treated as impersonal and inanimate things. Coral Ann Howells comments: "Atwood's feminist concerns are plain here but so too are her concerns for basic human rights" (Margaret Atwood 128). With tattoed ankles and an identity pass, the handmaids are just commodities. They exist as mere items in a system of contract and ownership.

In the Gileadean patriarchy, women are reduced to their organs; the handmaids are reduced to wombs. Ecofeminists are against this attitude of the patriarchy to isolate and reduce things on the basis of value and productivity. As Vandana Shiva states in *Ecofeminism*, the reductionist policy of science reduces complex ecosystems to a single component and to a single function. It manipulates the ecosystem in a way that maximises single function, single component exploitation. As Shiva states, "Reductionist science is at the root of the growing ecological crisis, because it entails a transformation of nature that destroys its organic processes and rhythms and regenerative capacities" (25). It creates the possibility of colonising and controlling that which is free and self-generative. Like the seed, the woman's body, as site of regenerative power has become a colony of the capitalist patriarchy. As Shiva points out, "These sites of creative regeneration are transformed into 'passive' site where the expert 'produces' and adds value" (25). Nature, women and non-white people provide the raw material for the capitalist patriarchy. In Gilead, woman is reduced to an incubator and man is elevated to that of a parent.

In the Gileadean regime, the handmaids are just reproductive prostitutes who have no control over their own body. The alternative to fertility is death, so biology becomes the destiny for the handmaids.

The individuality of the handmaids is totally erased and they are viewed as mere organs: "We are two-legged wombs, that's all: sacred vessels, ambulatory chalices" (HT 128). A handmaid is nothing but a "wandering womb" (HT 138), a passive receptacle. Bodily membranes are under the threat of male invasion. Penetration into women and that into nature are equated here.

The handmaids in Gilead are treated just as animals reared for the purpose of milk, meat, eggs and so on. So Offred says, "I wait, washed brushed, fed like a prize pig" (HT 120). Just as animals with suitable genes are chosen for selective breeding, the handmaids are selected to be breeders of children. Offred recognizes this function when she says. "We are for breeding purposes" (HT 128). When Ofwarren, formerly Janine, conceives, the wives comment on her: "A strong girl, good muscles. No Agent Orange in her family, we checked the records, you can never be too careful" (HT 109). Patriarchal culture, which is based on the dualistic concept of the 'self' and the 'other', links women, nature and animals and categorizes them as the 'other'. This sort of linkage serves to emphasise the inferiority of these categories.

An important theme that Atwood discusses in *The Handmaid's Tale* is the question of reproductory rights or the male control over female reproduction. Just as science with its specialised production methods, colonises seeds and makes the plant body the centre of experiments, the Gileadean patriarchy has colonised the womb of the handmaids. As Shiva says in *Ecofeminism*, "Colonization of seed, reflects the patterns of colonization of women's bodies. Profits and power become intimately linked to invasion into all biological organisms" (29). The patriarchy cleverly propagates the idea that it is the duty of women to beget children, in submission. In this process, the woman is completely objectified and alienated from her body.

Offred, at first, remains passive, but there exists in her a strong desire to endure and survive. The presence of Nick assures her a sense of security and with him, she feels that she is inside "a cave" (HT 253). Finally, Serena Joy finds out her secrets and she is accused of violating state policies. When the van comes for her, Nick says that it is 'Mayday', hinting the possibility of escape. But Offred does not seem to be confident of her future: "Whether this is my end or a new beginning I have no way of knowing. I have given myself over into the hands of strangers, because it can't be helped" (HT 276).

Woman is equated to Canada, threat to female body is equated to threat to Canada as a nation from America. Just as the bodily membranes of the women in Gilead face the threat of invasion, the borders of Canada are under the threat of neighbouring America. The imperialistic America refers to Canada sometimes as their frontyard, and sometimes as their backyard, both of which imply ownership. America is presented as the male aggressor, its masculinist qualities are epitomized in the Gileadean patriarchy, The Republic of Gilead is generally accepted as an incarnation of American fundamentalism of early 1980s. In an interview with Bonnie Lyons, Atwood gives the reason for setting the novel outside Canada:

> Because it's not a Canadian sort of thing to do. Canadians might do it after the States did it, in some sort of watered down version. The mind-set of Gilead is really close to that of the seventeenth -century Puritants. It is true that everyone watches the States to see what the country is doing and might be doing ten or fifteen years from now (223).

Fertile land and fertile woman are equally exploited by the capitalistic patriarchy. Exploited women and exploited land suffer the same impoverishment. The inhabitants of Gilead lead the precarious life of victims, trying to endure and survive. It is significant that Offred often compares herself to a space: 'I am like a room where things once happened and now nothing does (HT 98).

In *The Handmaid's Tale*, Atwood uses ecofeminist metaphors linking body with the earth, vegetation and organisms. Ecofeminist discourse attempts to expose the metaphors which keep women, nature and animals bound together, thereby revealing their collective subordination.

Atwood has subtly used many images relating body to the environment. Offred compares her own body to the earth: "I sink down into my body as into a swamp, fenland, where only I know the footing. Treacherous ground, my own territory. I become the earth, I set my ear against, for rumours of the future" (HT 69). She waits for a child with expectation, just as a seed waits to germinate. Women are associated with nature, through the garden imagery. Offred often indulges in memories of the past: "I once had a garden. I can remember the smell of the turned earth, the plump shapes of bulbs held in the hands, fullness, the dry rustle of seeds through the fingers" (HT 112). She feels that the scent from the garden "rises like heat from a body" (HT 179). It is significant that many wives resort to gardening to achieve inner autonomy.

In *The Handmaid's Tale*, there are many images linking female body with vegetation. As part of their indoctrination, Aunt Lydia says to the handmaids: "Think of yourselves as seeds [...]" (HT 18). Offred sees flowers as the "genital organs of plants" (HT 77). As Roberta Rubenstein points out, "distinctions between human and non-human are grotesquely inverted or reduced" ("Nature and Nurture" 107.) The body of nature is often attributed with female dimensions. The bleeding hearts that blossom in Serena Joy's garden are "female in shape" (HT 143). Offred, while walking after a rainy night observes that the gravel path that neatly divides the damp back lawn is like a "hair parting" (HT 17). Offred identifies herself with the vegetation: "Winter is not so dangerous. I need hardness, cool, rigidity; not this heaviness, as if I'm a melon on a stem, this liquid ripeness" (HT 144). When Offred suspects that Ofglen is a spy, she thinks: "It occurs to me that she may be a spy, a plant set to trap me; such is the soil in which we grow. But I can't believe it; hope is rising in me, like sap in a tree. Blood in a wound. We have made an opening" (HT 158).

In *The Handmaid's Tale* there are images linking body with various organisms. A variety of creatures—worm, oyster, spider, beetle, mouse, fish, frog, snake, hawk, vulture, chicken, pig, dog, cat, ant—are mentioned frequently in the narrative to denote the handmaids' parallel existence with them. The narrator Offred says about Ofglen, "She walks demurely head down, red-gloved hands clasped in front, with short little steps like a trained pig's on its hind legs" (HT 19). The hanged bodies of the victims of Gilead are said to be like meat on meat hooks. Such organic links occur frequently in the novel—it is said that from Janine, a "smell, more animal" is coming, "a smell of dens, of inhabited caves" (HT 116).

In *The Handmaid's Tale*, Atwood depicts how the indiscriminate application of science and technology can threaten the survival of woman and nature. Capitalistic patriarchy, with the assistance of science, having colonised the resources of nature, has begun to colonise the spaces within woman, vegetation and animals. Thus, in the dystopian society of Gilead, woman is reduced to a womb and is denied power over her own body. The colonised body becomes a metaphor for the colonised existence of Canada. The ecofeminist metaphors, which relate female body to nature and organisms, foreground their collective colonisation. In this novel, Atwood argues for a more just society where the rights of women and the laws of nature are well protected.

Works Cited

Atwood, Margaret. *The Handmaid's Tale.* 1985. Toronto: Seal Books, 1986.

____ *Second Words: Selected Critical Prose.* Toronto: Anansi, 1982.

____ Interview with Bonnie Lyons. "Using Other People's Dreadful Childhoods". Ingersoll, *Conversations*. Princeton: Ontario Review P, 1990.

Gaard, Greta, ed. "Living Interconnections with Animals and Nature". *Ecofeminism*: *Women, Animals, Nature* 1-12. Philadelphia: Temple UP, 1993.

Givner, Jessie. "Names, Faces and Signatures in Margaret Atwood's *Cat's Eye* and *The Handmaid's Tale. Canadian Literature.* 133 (1992): 56-75.

Howells, Coral Ann. *Margaret Atwood.* London: Macmillan P Ltd., 1996.

Mies Maria and Vandana Shiva. *Ecofeminism.* New Delhi: Kali for Women, 1993.

O' Loughlin Ellen. "Questioning Sour Grapes: Ecofeminism and the United Farm Worker's

Grape Boycott". Gaard, *Ecofeminism: Women, Animals and Nature* 146-166.

Rubenstein, Roberta. "Nature and Nurture in Dystopia: *The Handmaid's Tale*". Van spanckeren and Castro. Eds. *Margaret Atwood: Vision and Forms*. Carbondale: Southern Illinois UP, 1988.

12

Ecocritical Reading of Yann Martel's *Life of Pi*

— S.M. Gupta

Ecocriticism is the study of interconnectedness between man and environment mediated through gender, class, race, science, culture and economics.

The novel *Life of Pi* demonstrates this interrelationship in several ways. The novel is set mainly on the Pacific Ocean. The hero of the novel has been named after a Parisian swimming club called Piscine Molitor (Patel); before the shipwreck he lives in the zoo situated in Pondicherry; his father is a zoo keeper; he studies zoology later for his B.S. Honours at the University of Toronto.

He holds the view that a zoo is a place of 'security' and 'plenty' for animals. The descriptions of the animal behavior, flight distances, dislike for invasion on their (animal's) territory, zoomorphism and cruelty of man against zoo animals are very insightful. The whole description can be termed as 'animalus anthropomorphicus'.

He expounds the notion of interconnectedness through the natural cycle of 'food chain'. The hyena eats the zebra and the orangutan on the life-boat and in turn she is eaten by Richard Parker, the tiger. Mr. Pi spends 227 days with a ferocious tiger, on a rough and stormy ocean and a carnivorous island without potable water and food but learns how to survive in adverse environment. This underscores man's integration with nature.

The novel contains vivid and pictorial descriptions of seascape, marine life, dolphins, birds, albatrosses, Wilson's petrels, a short tailed shearwater, masked boobies; vivid descriptions of weather, rain, storms, blasts, thunder, lightning and a carnivorous island—'a floating organism of algae of leviathan proportions.'

The order of this eco-system in and on the sea is spoiled by man who pours tons of waste materials on the sea shores surrounding big towns and metropolises.

The whole novel underlines the relationship and interconnectedness between man and nature and a constant epic struggle between man and nature. But finally it is shown that by striking a balance between man and nature, man can ensure his survival in adverse environment.

Introduction

I would like to begin this article with an excerpt from a poem' Give and Take' by Roger McGough, a leading British poet, whose poems have been included in Feeling the Pressure, The Poetry and Science of Climate Change ('Connecting', April-May 2008:21)

I give you clean air
you give me poisonous gas
I give you mountains
you give me quarries
I give you pure snow
you give me acid rain
I give you spring fountains
you give me toxic canals.

—(Roger McGough 21)[1]

The other excerpt is from *Times of India* (19 May 2008) which is titled as: 'Twelve Species on the Brink of Extinction and the Bare Truth.'

These two excerpts highlight two different truths about man's relationship with nature. The first one demonstrates how man has defiled and is defiling nature whereas the second excerpt shows what animal lovers are doing across the globe to undo the cruel treatment meted out to the animals through the organization of 'Peta' (People for the Ethical Treatment of Animals). It also shows that some rare species of animals are on the brink of extinction and if nothing is done to salvage these species, they will be lost for ever.[2]

This division between destroyers of environment and lovers of environment is pretty old. One is reminded of Amrita Devi of Jodhpur. In the eighteenth century the then king of Jodhpur wanted to construct a new palace on a hilly terrain. He ordered the woodcutters to cut all the trees situated there. Amrita Devi of Jodhpur embraced the tree that they wanted to cut. Later on, her family members also joined her and thus a small protest took the form of a movement. In 1970s, this movement came to be known as 'chipko aandolan' spearheaded by Chandi Prasad Bhatt and Sundarlal Bahuguna. These people want to protect environment so that there is a balance between man and nature and this balance will ensure the well- being of future generations.

Ecofeminists Vandana Shiva, Wangari Maathai and Karen J. Warren link the destruction of environment with the domination and exploitation of women by men. For over three decades, Wangari Maathai has been advocating the rights of forests and women, the rights of the easily overlooked "Other".

It would not be inappropriate to say a few words about ecology, environment, eco-criticism and the goal of eco-criticism at this stage. It should be noted that terms like 'enviro-' have been rejected because they smack of anthropocentricism and terms like 'eco-' have been accepted because they show interdependence between human and non-human.

Ernst Heirich Hackel (1834-1919), a German biologist introduced the term 'ecology' for the first time. He applied the terms 'oekologie 'to the relation of the animal both to its organic as well as its inorganic environment.

Thus ecology deals with the relations and interactions between organisms and their environment including other organisms. Nature writing has been replaced by ecological literature. Since 1960s ecology has emerged as a significant term.

According to Bhanja (2005), in Sanskrit the word nature refers to *Brahma-ananda*, which includes *Panchamahabutani* (earth, water, fire, sky, air), three *Vyahtritis* or mystical syllables *bhur* (the earth) + *bhuvah* (either) + *svah* (heaven). Sometimes it is called *Prakriti* (primal matter or substance) from which physical and mental universe evolves under the influence of *Purusha* (one's true self).

He further states that Nature has always been accorded the status of the Great Mother since Pre-Christian or Vedic times. In Christianity there is an implication that nature is created for human beings who are her children and thus have a right to exploit nature/ecology. In Darwin's *Origin of Species* (1859) nature is portrayed as essentially 'hostile' and 'baleful' and 'apathetic' to human beings. Thus religion and science have rationalized the dominance of man over nature alias ecology.

The term 'ecocriticism' was coined in 1970 and became a movement in 1990s. In 1992 a new association (ASLE) i.e. Association for the Study of Literature and Environment was formed under the leadership of Scott Slovic. The purpose of this association was to study relations between human beings and the natural world; and to encourage "new nature writing, traditional and innovative; scholarly approaches to environmental literature and an interdisciplinary ecological research."

Ecocriticism is the study of relationship between literature and environment, that is application of ecological concepts to the study of literature. Glotfelty (1996), Evan Mwangi, Ursula K. Heise, Simon C. Estok, Christopher Cokins, Thomas K. Dean, Allison B. Wallace, David W. Teague, David Taylor, Mark Schlenz, Stephanie Sarver, Kent Ryden, Ian Marshall, Joan Hamilton, Don Scheese, Ralph W. Black, Stan Tag, Nancy Cook and a number of other scholars in ASLE'S 'related conferences and abstracts' and in 'introduction to ecocriticism' have discussed the theory of ecocriticism.[3] Some of their views are summarized below. Nature should be viewed not as a commodity but as part of our community (Joan Hamilton); Lawrence Buell (1996:430) defines "ecocriticism as study of the relationship between literature and the environment conducted in a sprit of commitment to environmental praxis."[4]

According to Heise, (see internet reference n.10), 'some ecocritics understand their intellectual work as a direct intervention in current social, political and economic debates surrounding environmental pollution and preservation'. Cokinos (vide n.10) is of the view that 'eco-criticism is the critical and pedagogical broadening of literary studies to include texts that deal with the nonhuman world and our relationship to it'. According to Dean (ibid), 'ecocriticism is a study of culture and cultural products (art works, writings, scientific themes etc.) that is in some way connected with the human relationship to the nature world'. Wallace (ibid) opines that 'Both eco literature and eco criticism address and redress this general American failure (ecological crisis) to participate as fully—that is as imaginatively as possible. In all the nonhuman' (underlining mine). Alexander Wilson (quoted in David Taylor, vide internet reference in n.10) claims that "the culture of nature—the ways we think, teach, talk about, and construct the natural world—is as important terrain as the land itself."

According to Mark Schlenz, in ecocriticism, "human and natural resource exploitations are invariably linked through various and competing ideas about nature" (ibid).

According to Stephanie Sarver (ibid), an eco-critic, using different literary theories considers how nature is reflected and perceived in literary texts. We should be able to demonstrate that environmental issues are human issues.

Kend Ryden (ibid) believes that "the ecocritical stance reconnects literary study to both the processes and the problems inherent in living on this heavily burdened planet, focusing our attention anew on the ground beneath our feet, on our complex relationships to that ground..."

In the preceding paragraphs, I have tried to acquaint the readers with a number of views of various critics on ecocriticism. As we can see, all of them have emphasized that the major concern of ecocriticism is the interconnectedness between human and non-human world as perceived and reflected in a literary text. Ecocritical theory is not a theory per se but it can take to any theoretical framework say postcolonial, post structural, feminist etc. to discuss the interconnectedness between man and ecology.

In this article, Yann Martel's *Life of Pi* will be discussed from the point of view of eco-criticism. The interconnectedness between man and nature will be indicated through the natural cycle of 'food-chain'. The protagonist is ship-wrecked and is compelled to spend 227 days on a life-boat in the Pacific Ocean-with wild and ferocious animals. The protagonist learns to survive in adverse environment. This highlights man's integration with nature. The novel also contains beautiful descriptions of marine life. The eco-system is polluted by man's pouring of tons of waste material into the sea. The message derived from the novel is that by striking a balance between man and nature, man can survive in adverse ecological circumstances. All those aspects of ecological interpretation will be discussed in great detail in the following pages.

Interconnectedness

The novel *Life of Pi* shows the interconnectedness between man and nature in several ways. The part two of the novel is set on the Pacific Ocean. The protagonist has been named after a Parisian swimming club called Piscine Molitor (Patel). It is reminiscent of ecosystem. Before the shipwreck, he lives in the zoo situated in Pondicherry and his father is a zoo-keeper.

He states that 'portion of the grounds of the Pondicherry Botanical Garden was made available rent free for an exciting business opportunity and... India had a brand new zoo... (*Life of Pi*: 12) Pi's father, Mr. Santosh Patel, who was a hotelier in Madras, entered the zoo business because of his 'abiding interest in animals'. Describing the zoo Pi states:

> It was a huge zoo, spread over numberless acres, big enough to require a train to explore it... the riot of flowers is incessant. There are trees, shrubs and climbing plants in profusion—peepuls, gulmohors, flames of the forest, red silk cotton, jacarandas, mangoes, jack fruits and many others that would remain unknown to you if they didn't have neat labels at their feet. ... And when you turn your head you see the elephant that was there all along, so big you didn't notice it. And in the pond you realize those are hippopotamuses floating in the water. The more you look the more you see. You are in zootown. (*Life of Pi*. 12-13)

Mr. Pi has sweet memories of growing up in a zoo in the enviable company of zoo animals such as lions, monkeys, mynahs, Moluccan cockatoos, American bisons, yawning orangutans, shiny frogs and a number of species of birds such as pink flamingoes or black swans, or one—wattled cassowaries, silver diamond doves, cape glossy starlings, peach-faced love-birds, Nanday conures and orange fronted parakeets. There are also seals, the big cats, bears, baboons, macaques, mangabeys, gibbons, deer, tapirs, llamas, giraffes and mongooses, were early risers.

Visiting a zoo, according to Pi, at sunrise and sunset accords the greatest sensory pleasure:

> In zoos, as in nature, the best times to visit are sunset and sunrise. That is when most animals come to life. They stir and leave their shelter and tip-toe to the water's edge. They show their raiments. They sing their songs. They turn to each other and perform their rites. The reward for the watching eye and the listening ear is great. I spent more hours than I can count a quiet witness to the highly mannered, manifold expressions of life that grace our planet. It is smell: so bright, loud, weird and delicate as to stupefy the senses. (*Life of Pi:* 15)

Pi believes that people have a wrong notion about zoos as they have about God and religion. They think that animals are happier and freer in the wild than in the zoo. Pi is of the view that like human beings animals are 'territorial'. 'Only a familiar territory will allow them to fulfill the two relentless imperatives of the wild: the avoidance of enemies and getting the food and water'. 'A house is a compressed territory where our basic needs can be fulfilled close by and safely. In a zoo, an animal will feel like a 'landholder'. A zoo is a place of 'plenty' and safety for animals. They feel 'happy', contended and relaxed in a zoo. The animal lovers may not buy this idea.

Pi makes insightful comments on the behavior of man towards animals. He states that we commonly say in the trade that the most dangerous animal in the zoo is man' (*LoP*: 29). Man is predatory by nature and through this quality has made the entire planet his prey. The most burning question today is the cruelty of man towards flora and fauna. Pi gives graphic details of this cruelty in the following excerpt.

> ...we have in mind the people who feed fishhooks to the otters, razors to the bears, apples with small nails in them to the elephants and hardware variations on the theme: ballpoint pens, paper clips, safety pins, rubber bands, combs, coffee spoons, brooders, and other jewellery (and not just cheap plastic bangles gold wedding bands too,) drinking straws, plastic and so on. The obituary of zoo animals that have died being fed foreign bodies would include gorillas, bison, storks, rheas, ostriches, seals, sea lions, big cats, bears, camels, elephants monkeys, and most every variety of deer, ruminant and songbird. Among zookeepers, Goliath's death is famous, he was a bull elephant seal, a great big venerable beast of two tons, star of his European zoo, loved by all visitors. He died of internal bleeding after someone fed him a broken beer bottle. (*LoP*: 29)

The list is endless. Pi states that human beings have the tendency to see animals through their eyes (*animalus anthropomorphicus*).

'The obsession with putting ourselves at the centre of everything is the bane not only of theologians but also of zoologists. (*LoP*: 31). Pi's father one day takes Pi and his brother through the zoo and cautions them that animals except domesticated animals are dangerous. Tigers are dangerous and so are lions, hippos, hyenas, bears, orangutan, Arabian camels, elephants, black swans. Except guinea pigs, other animals are dangerous. These comments are an example of *animalus anthropomorphicus.*

Pi describes flight distances of animals. Flight distance is the minimum distance at which an animal wants to keep a perceived enemy. 'A flamingo in the wild won't mind you if you stay more than three hundred yards away. Cross that limit and it becomes tense. Different animals have different flight distances and they gauge them in different ways. Cats look, deer listen, bears smell. He further states.

> Giraffes will allow you to come to within thirty yards of them if you are in a motor car, but will run if you are 150 yards away on foot. Fiddler crabs scurry when you're ten

> yards away; howler monkeys stir in their branches when you're at twenty; African buffaloes react at seventy five. (*LoP*: 39).

Any animal would escape if not kept in a particular habitat. Animals do not escape to somewhere but from something. Animals hate the unknown. Escaped zoo animals are not dangerous absconding criminals but simply wild creatures seeking to fit in. An animal would never like his territory to be invaded. Pi rightly observes:

> ... If you fall into a lion's pit the reason the lion will tear you to pieces is not because it's hungry—be assured, zoo animals are amply fed- or because it's bloodthirsty, but because you've invaded its territory (*LoP*: 43).

Zoo-keeping is a brilliant example of interrelationship between man and animals. Visiting zoos emphasizes our love for animals. Watching animal behavior, habitat and their flight distances, territoriality—all show that we want to live in peaceful co-existence with fauna in the ecosystem.

Recall the definition of ecocriticism to the effect that it is the study of interconnectedness between man and environment mediated through class, gender, race, science, culture and economics. In zoo keeping, it is economics that interacts between the relationship of man and animals. Man imprisons animals in a zoo, tearing them away form their natural habitat to earn money. It is a business tactics to enslave animals and cannot be called an ethical act at all. Zoo visitors (most of them though) go to the zoo out of curiosity; they want to see the animals and study their behavior. As described by Pi that some of them perform cruelty towards them, which is not pardonable.

They argument of Pi that zoo is a place of plenty and security is specious. Doesn't it show man's assertion of supremacy and control of power over the powerless animals by scheming man? He treats them the other and enslaves them for his entertainment.

This approach of man instead of creating a balance in environment creates a fissure and cannot be defended in any case. Further, all animals except guinea pigs have been described as dangerous by Pi's father. This anthropomorphous view of animals is a part of man's assertion of power over animals who are a very important part of ecosystem.

Food-Chain

He expounds the notion of inter connectedness between various objects of nature through the natural cycle of 'food-chain'. A food chain is the flow of energy from one organism to the next. Organisms in a food chain are grouped into topic levels and mash, based on how many levels they are removed form the primary producers. They usually start with primary producers and end with a carnivore. The base of food-chain is phytoplankton, by the virtue of their ability to photo-synthesize. The food chain shows only one pathway of energy and material transfer. Most consumers feed on multiple species and are in turn fed upon by multiple other species. The food chain has a producer, consumer, herbivore, carnivore, and 'decomposer.[5] There are three types of consumers: primary, secondary and tertiary. Eventually this transfer of energy ends with decomposers (dead animals) that are broken down and used as food or nutrition by bacteria or fungi. Pi describes an interesting case in the zoo in which the predator- prey relationship was temporarily suspended but ultimately it was restored proclaiming the continuation of food-chain:

> We had our own case of the freak suspension of the predator-prey relationship. We had a mouse that lived for several weeks with the vipers. While other mice dropped in the terrarium disappeared within two days, this little brown Methuselah built itself a nest, stored the grains we gave it in various hideaways and scampered about in plain sight of the snakes. We were amazed. We put up a sign to bring the mouse to the public's attention. It finally met its end in a curious way: a young viper bit it? Was the viper unaware of the mouse's special status? Unsocialized to it, perhaps? Whatever the case, the mouse was bitten by a young viper but devoured—and immediately by an adult. (underlining mine).If there was a spell, it was broken by the young one. Things returned to normal after that. All mice disappeared down the vipers' gullets at the usual rate. (*LoP*: 85).

Pi, travelling by ship named as Tsimtsum to Canada with his family and a few zoo animals, realized that the ship was sinking. He found it as 'unbelievable as the moon catching fire'. He found himself all alone and orphaned in the middle of the Pacific Ocean, hanging to an oar, an adult tiger in front of him sharks beneath him, a storm him raging about. He was in the lifeboat and along with him were Richard Parker (a tiger) a zebra, a hyena and the Orange Juice, a female orangutan.

After thirty-six hours, the hyena attacks the zebra:

> Then, with only a snarl for notice, the hyena went amok. It hadn't moved form its cramped quarters all day. It put its front legs on the zebra's, reached over and gathered a fold of skin in its jaws. It pulled roughly. A strip of hide came off the zebra's belly like gift-wrap paper comes off a gift, in a smooth-edged swath, only silently, in the way of tearing skin, and with greater resistance. Immediately blood poured forth like a river. Barking, snoring and squealing, the zebra came to life to defend itself. It pushed on its front legs and reared its head in an attempt to bite the hyena, but the beast was out of reach. It shook its good hind leg, which did no more than—explain the origin of the previous night's knocking: it was the hoof beating against the side of the boat. The zebra's attempts at self-preservation only whipped the hyena into a frenzy of snarling and biting. It made a gaping wound in the zebra's side. When it was no longer satisfied with the reach it had from behind the zebra, the hyena climbed onto its haunches. It started pulling out coils of intestines and other viscera. There was no order to what it was doing. It bit here, swallowed there seemingly overwhelmed by the riches before it. After devouring half the liver, it started tugging on the whitish, balloon-like stomach bag. But it was heavy, and with the zebra's haunches being higher than its belly—and blood being slippery—the hyena started to slide into its victim. It plunged head and shoulders into the zebra's guts, up to the knees of its front legs. It pushed itself out, only to slide back down. It finally settled in this position, half in, half out. The zebra was being eaten alive from the inside. (*LoP*: 125).

The graphic description of the eating away of the zebra by the hyena is an example of the food-chain found in nature.

> In the afternoon, violence broke out once again. The hyena was yipping. Orange Juice was grunting and making loud lip- smacking noises. All of a sudden their complaining fused and shot up to top volume. The hyena jumped over the remains of the zebra and made for Orange Juice. She thumped the beast on the head. It was something shocking. The orangutan defended herself well in the beginning.
>
> An adult female orangutan cannot defeat an adult male spotted hyena. That is the plain empirical truth. Let it become known among zoologists. Had Orange Juice been a male, had she loomed as large on the scales as she did in my heart, it might have been another matter. But it is not simply a question of weight and brute strength. Orange Juice was far from defenceless. What it comes down to is attitude and knowledge. What does a fruit eater know about killing? Where would it learn where to bite, how hard, for how long? An orangutan may be taller, may have very strong and agile arms and long canines, but if it does not know how to use these as weapons, they are of little use. The hyena with only its jaws will overcome the ape because it knows what it wants and how to get it.
>
> The hyena came back. It jumped on the bench and caught Orange Juice at the wrist before she could strike. Orange Juice hit the hyena on the head with her other arm, but the blow only made the beast snarl viciously. She made to bite, but the hyena moved faster. Alas, Orange Juice's defence lacked precision and coherence. Her fear was something useless that only hampered her. The hyena let go of her wrist and expertly got to her throat.
>
> Dumb with pain and horror, I watched as Orange Juice thumped the hyena ineffectually and pulled at its hair while her throat was being squeezed by its jaws. To the end she reminded me of us: her eyes expressed fear in such a humanlike way as did her strained whimpers. She made an attempt to climb onto the tarpaulin. The hyena violently shook her. She fell off the bench to the bottom of the lifeboat, the hyena with her. I heard noises but no longer saw anything. (*LoP*: 130-131)

It is often said that *jivo jivasya bhojanam* (one creature is another creature's food). Darwin states the Principle of Survival of the Fittest. Strength counts; but then nature has endowed every creature with a mechanism through which it can get itself food and survives. The hyena in this case is the predator and the orangutan is the prey. The cycle of food-chain continues in nature creating an ecological balance.

In the lifeboat are left Richard Parker (the Bengal tiger), the hyena and Pi. Richard Parker remained passive for three days most probably under the effect of sedation and seasickness. Pi was aware of the reality of the tiger. But Richard Parker before attempting to attack Pi, finished the hyena:

> I saw a sight that will stay with me for the rest of my days. Richard Parker had risen and emerged. He was not fifteen feet from me. Oh, the size of him! The hyena's end had come, and mine. I stood rooted to the spot, paralyzed in thrall to the action before my eyes. My brief experience with the relations of unconfined wild animals in lifeboats had made me expect great noise and protest when the time came for bloodshed. But it happened practically in silence. The hyena died neither whining nor whimpering, and Richard Parker killed without a sound. The flame-coloured carnivore emerged from beneath the tarpaulin and made for the hyena. The hyena was leaning against the stern bench, behind the zebra's carcass, transfixed. It did not put up a fight. Instead it shrank to the floor, lifting a forepaw in a futile gesture of defence. The look on its face was of terror. A massive paw landed on its shoulders. Richard Parker's jaws closed on the side of the hyena's neck. Its glazed eyes widened. There was a noise of organic crunching as windpipe and spinal cord was crushed. The hyena shook. Its eyes were dull. It was over. (*LoP*: 150-151)

Notice the end of the hyena is not that cruel. It looks quite cool. But the end of the hyena is part of the food chain.

Now in the lifeboat are left Richard Parker and Pi. Pi has to figure out how to save himself from this ferocious animal otherwise he would be the next prey. But as we know man is the most intelligent person among all the creatures in the world, Pi makes several plans to kill the tiger. Ironically, it was Richard Parker who calmed Pi down. So Pi also decides to keep

Richard Parker alive. Ruminating on the past happenings, Pi states:

> Now we were two. In five days the populations of orangutans, zebras, hyenas, rats, flies and cockroaches had been wiped out. Except for the bacteria and worms that might still be alive in the remains of the animals, there was no other life left on the lifeboat but Richard Parker and me. (*LoP*: 170-171)

Pi, very intelligently drew a separate territory for himself and Richard Parker and found ways not to let Parker enter his territory. Both Pi and Richard Parker live on different kind of sea-life for their sustenance:

> Richard Parker turned and started clawing the shark's head with his free front paw and biting it with his jaws, while his rear legs began tearing at its stomach and back. The shark held on to his paw, its only line of defence and attack, and thrashed its tail. Tiger and shark twisted and tumbled about. With great effort I managed to gain enough control of my body to get into the raft and release it. The lifeboat drifted away. I saw flashes of orange and deep blue, of fur and skin, as the lifeboat rocked from side to side. Richard Parker's snarling was simply terrifying. (*LoP*: 220)

After the hardship of a number of days, Pi had a stroke of good luck when he hit upon ponds of fresh water. He touched an island which was six or seven miles in diameter. This island was not an island in the conventional sense of the term- that is a small hand mass rooted to the floor of the ocean- but was rather a free- floating organism a ball of algae of leviathan proportions. On this strange island of trees that grow directly out of vegetation without any soil, Pi and Richard Parker stay here for a time, sleeping in their boat and exploring the island during the day. Pi discovers a huge colony of meerkats who sleep in the trees and freshwater ponds. One day, Pi finds human teeth in a trees fruit and comes to the conclusion that the island eats people. He and Richard Parker head back out to sea, finally reaching a Mexican beach. Richard Parker runs off, and villagers take Pi to hospital. Pi describes the carnivorous island in the following words:

> The island was carnivorous. This explained the disappearance of the fish in the pond. The island attracted salt water fish into its subterranean tunnels- how I don't know; perhaps fish ate the algae as gluttonously as I did. They became trapped. Did they lose their way? Did the openings onto the sea close off? Did the water change salinity so subtly that it was too late by the time the fish realized it? Whatever the case, they found themselves trapped in fresh water and died. Some floated up to the surface of the ponds, the scraps that fed the meerkats. At night, by some chemical process unknown to me but obviously inhibited by sunlight, the predatory algae turned highly acidic and the ponds became vats of acid that digested the fish. This was why Richard Parker returned the boat every might. This was why the meerkats slept in the trees. This was why I had never seen anything but algae on the island. (*LoP*: 281:282)

The description of this carnivorous algae is another illustrious example of food chain in nature. The description a murderous island is not a piece of fantasy. Science Reporter (July 2008) discusses such strange plants called 'bladderwort' (a water plant) which sits in the water and traps its prey and then proceeds to eat it.[6]

Space for Greater Bladderwort

Order in Nature

There are vivid and pictorial descriptions of seascape, marine life, dolphins, sharks, birds, albatrosses (compare *The Ancient Mariner*) Wilson's petrels; vivid description of weather, rain, storms, blasts, thunder, lightning and a carnivorous island in the novel. All this multiplicity of flora and fauna in the novel highlight the order in nature and interconnectedness between man and nature. Nature in its various threatening aspects (blasts, thunder, lightning, storms etc.) plays a very significant role in awe-inspiring setting of the novel.

Pi and his family are going to Canada by ship. He takes a look at the 'dangerous, beautiful and forbidding sea standing on the deck of the ship:

> ...As for the sea, it looked rough, but to a landlubber the sea is always impressive and forbidding, beautiful and dangerous. Waves were reaching up, and their while foam, caught by the wind, was being whipped against the side of the ship. (*LoP*: 102)

Nature can be benevolent as well as malevolent. The flying fish hit Pi and Richard Parker like missiles; like arrows:

> It was fish. There was a fish in the locker. It was about 15 inches long and it had wings. A flying fish. Slim and dark grey-blue, with dry, featherless wings and round, unblinking yellowish eyes. It was this flying fish that struck me across the face, not Richard Parker. He was still fifteen feet away, no doubt wondering what I was going on about. But he had seen the fish. I could read a keen curiosity on his face. He seemed about ready to investigate.
>
> (*LoP*: 180:181)

There is a vivid description of the marine life in the novel. The sight of the marine life provided relief to Pi. This underscores the interconnectedness between man and nature:

> The underside of the raft became host to a multitude of sea life, like the net but smaller in form. It started with a soft green algae that clung to the life jackets. Sciffer algae of a darker kind joined it. They did well and became thick. Animal life appeared. The first that I saw were tiny, translucent shrimp, hardly half an inch long. They were followed by fish no bigger that looked like they were permanently under X-ray; their internal organs showed through their transparent skins. After that I noticed the black worms with the white spines, the green gelatinous slugs with the primitive limbs, the inch-long, motley-coloured fish with the potbellies, and lastly the crabs, half to three-quarters of an inch across and brown in colour. I tried everything but the worms, including the algae. Only the crabs didn't have an unpalatably bitter or salty taste. Every time they appeared, I popped them one after another into my mouth like candy until there were none left. I couldn't control myself. It was always a long wait between fresh crops of crabs.

> The hull of the lifeboat invited life too; in the form of small gooseneck barnacles. I sucked their fluid. Their flash made for good fishing bait.
>
> I became attached to these oceanic hitchhikers, though they weighed the raft down a little. They provided distraction like Richard Parker. I spent many hours doing nothing but lying on my side, a jacket pushed out of place a few inches, like a curtain from a window, so that I might have a clear view. What I saw was an upside-down town, small, quiet and peaceable, whose citizens went about with the sweet civility of angels. The sight was a welcome relief for my frayed nerves. (*LoP*: 197-198)

This description of nature continuous in the novel. Nature is sometimes baleful, sometimes benevolent. Struggling with different moods of nature, Pi too experiences different moods ranging from fear, rage, madness, hopelessness to apathy. Consider the following description of the skies and seas:

> There were many skies. The sky was invaded by great white clouds, flat on the bottom but round and billowy on the top. The sky was completely cloudless, of a blue quite shattering to the senses. The sky was a heavy, suffocating blanket of grey cloud, but without promise of rain. The sky was thinly overcast. The sky was dappled with small, white fleecy clouds. The sky was a featureless milky haze. The sky was a density of dark and blustery rain clouds that passed by without delivering rain. The sky was painted with a small number of flat clouds that looked like sandbars. The sky was a mere block to allow a visual effect on the horizon: sunlight flooding the ocean, the vertical edges between light and shadow perfectly distinct. The sky was a distant black curtain of falling rain. The sky was many clouds at many levels, some thick and opaque, others looking like smoke. The sky was black and spitting rain on my smiling face. The sky was nothing but falling water, a ceaseless deluge that wrinkled and blocked my skin and froze me stiff.
>
> There were many seas. The sea roared like a tiger. The sea whispered in your ear like a friend telling you secrets. The sea clinked like small change in a pocket. The sea thundered like avalanches. The sea hissed like sandpaper working on wood. The sea sounded like some one vomiting. The sea was dead silent. And in between the two in

> between the sky and the sea were all the winds. And there were all the night and the moons.
>
> To be a castaway is to be a point perpetually at the centre of a circle. However, much things may appear to change—the sea may shift from whisper to rage, the sky might go from fresh blue to blinding white to darkest black--the geometry never changes. Your gaze is always a radius. The circumference is ever great. In fact, the circles multiply. To be castaway is to be caught in a harrowing ballet of circles. You are at the centre of one circle, while above you two opposing circles spin about. The sun distresses you like a crowd, a noisy, invasive crowd that makes you cup your ears, that makes you close your eyes, that makes you want to hide. The moon distresses you by silently reminding you of your solitude; you open your eyes wide to escape your loneliness. When you look up, you sometimes wonder if at the centre of a solar storm, if in the middle of the Sea of Tranquility, there isn't another one like you also looking up, also trapped by geometry, also struggling with fear, rage, madness, hopelessness, apathy. (*LoP*: 215-216)

The passages Abound in the Freshest Comparisons

Yann Martel not only describes the natural phenomena such as rain, seas, and the moon, but also the fauna that is found in the sea and which keep coming to the sea. Pi establishes a kind of relationship with the fauna while being in the life-boat. Pi finds whales very sensitive and responsive to his condition. There develops a bond between him and the whales. He views the birds with envy because they are free but he cannot get away from the sea.

> We saw a number of whales but none so close up as the first one, I would be alerted to their presence by their spouting. They would emerge a short distance away, sometimes three or four of them, a short-lived archipelago of volcanic islands. These gentle behemoths always lifted my spirits. I was convinced that they understood my condition, that at the sight of me one of them exclaimed, "Oh! Its' that castaway with the Pussy cat Bamphoo was telling me about. Poor boy. Hope he had enough plankton. I must tell Mumphoo and Tomphoo and Stimphoo about

> him. I wonder if there isn't a ship around I could alert. His mother would be happy to see him again. Goodbye, my boy. I'll try to help. My name's Pimphoo. (*LoP*: 230)

Pi also washed the fun-loving dolphins with pleasure:

> Dolphins were fairly regular visitors. One group stayed with us a whole day and night. They were very gay. Their plunging and turning and racing just beneath the hull seemed to have no purpose other than sporting fun. I tried to catch one. But none came close to the gaff. And even if one had, they were too fast and too big. I gave up and just watched them. (*LoP*: 230)

Pi found the birds very attractive. He was amazed to see them. The albatross is incomprehensible for him:

> I saw six birds in all. I took one to be an angel announcing nearby land. But these were sea---- fearing birds that could span the pacific with hardly a flutter of the wings. I watched them with awe, and envy and self pity. (*LoP*: 230)
>
> Twice I saw an albatross. Each flew by high in the air without taking any notice of us. I started with my mouth open. They were something supernatural and incomprehensible. (*LoP*: 230)

The birds, unlike whales, are quite apathetic to the condition of Pi:

> Another time, a short distance from the boat, two Wilson's petrels skimmed by, feet something on the water. They, too, took no notice of us, and left me similarly amazed. (*LoP*: 230)

Pi at last succeeded in attracting the attention of a short-tailed shearwater circling above them. Pi tried to catch it but with no result. At last Pi succeeded in catching a masked booby. The description of the bird is quite fresh:

> It was a large bird with a pure snowy white body and wings that were jet-black at their tips and rear edges. Its big bulbous head had a very pointed orange-yellow beak and the red eyes behind the black mark made it look like a thief who had a very long night. (*LoP*: 231)

Nature presents itself in various aspects; some are friendly; others are dangerous and pose a threat to mankind. This threatening aspect of nature creates awe in man and this helps restore a balance between man and nature. Lightning, storm, bolt, heat, are the frightening aspects of nature.

Once there was lightning. The sky was so black, day looked like night. The downpour was heavy. I heard thunder far away. I thought it would stay at that. But a wind came up, throwing the rain this way and that. Right after, a white splinter came crashing down from the sky, puncturing the water. It was some distance form the lifeboat, but the effect was perfectly visible. The water was shot through with what looked like white roots; briefly, a great celestial tree stood in the ocean. I had never imagined such a thing possible, lightning striking the sea. The clap of thunder was tremendous. The flash of light was incredibly vivid.

> I turned to Richard Parker and said, "Look Richard Parker, a bolt of lightning." I saw how he felt about it. He was flat on the floor of the boat, limbs splayed and visibly trembling. (underlining mine *LoP*: 232)

When the bolt of lightning struck, it was deafening. Pi was dazed, thunderstruck:

> Suddenly a bolt struck much closer. Perhaps it was meant for us: we had just fallen off the crest of a swell and were sinking down its back when its top was hit. There was an explosion of hot air and hot water. For two perhaps three seconds, a gigantic, blinding white shard of glass from a broken cosmic window danced in the sky, insubstantial yet overwhelmingly powerful. Ten thousand trumpets and twenty thousand drums could not have made as much noise as that of bolt of lightning, it was positively deafening. The sea turned white and all colour disappeared. Everything was either pure white light or black shadow. The light did not seem to illuminate so much as to penetrate. As quickly as it had appeared, the bolt vanished—the spray of hot water had not finished landing upon us and already it was gone. The punished swell returned to black and rolled on indifferently. (underlining mine *LoP*: 233)

There was infernal heat in the atmosphere. This burnt his skin. His shrivelled muscles started aching:

> I clung to life. I was weakly frantic. The heat was infernal. I had so little strength I could no longer stand. My lips were hard and cracked. My mouth was dry and pasty, coated with a glutinous and saliva as foul to taste as it was to smell. My skin was burnt. My shrivelled muscles ached. My limbs, especially my feet, were swollen and a constant source of pain. I was hungry and once again there was no food. (*LoP*: 241)

Thus we see that there is an order in nature. But this order of the ecosystem is spoiled by man who pours tons of waste materials on the sea shores surrounding big towns and metropolises. An imbalance is created in nature causing and hinting at man's impending destruction:

> One day we came upon trash. First the water glistened with patches of oil. Coming up soon after was the domestic and industrial waste: Mainly plastic refuse in a variety of forms and colours but also pieces of lumber, beer cans, wine bottles, tatters of cloth, bits of rope and surrounding it all, yellow form. We advanced into it. I looked to see if there was anything that might be of use to us. I picked out an empty corked wine bottle. The lifeboat bumped into a refrigerator that lost its motor. It floated with its door to the sky. I reached out, grabbed the handle and lifted the door open. A smell leapt out so pungent and disgusting that it seemed to colour the air. Hand to my mouth, I looked in. There were stains, dark juices, a quantity of completely rotten vegetables, milk so curdled and infected it was a greenish jelly, and the quartered remains of a dead animal in such an advanced state of black putrefaction that I couldn't identify it. Judging by its size I think that it was lamb. In the closed, humid confines of the refrigerator, the smell had had the time to develop, to ferment, to grow bitter and angry. It assaulted my senses with a pent-up rage that made my head reel, my stomach churn and my legs wobble. Luckily, the sea quickly filled the horrid hole and the thing sank beneath the surface. The space left vacant by the departed refrigerator was filled by other trash.

> We left the trash behind. For a long time, when the wind came from that direction, I could still smell it. It took the sea a day to wash off the oily smears from the sides of the lifeboat. (*LoP*: 237)

Epic Struggle Survival and Balance between Man and Nature

Life of Pi is a story of epic struggle between man and nature. It is the story of epic struggle between Pi and different objects and phenomena of nature like the old man and the fish in the '*Old Man and the Sea*'. Pi is stranded on the Pacific Ocean for 227 days in the company of three ferocious animals; a hyena, an orangutan and a tiger. The ship by which Pi was to move to Canada had to move in adverse circumstances, though quite oblivious of the outside scenario:

> I don't understand. For days the ship had pushed on, bullishly indifferent to its surroundings. The sun shone, rain fell, winds blew, currents flowed, the sea built up hills, the sea dug up valleys—the Tsimtsum did not care. It moved with the slow massive confidence of a continent. (*LoP*: 100)

The sea looked rough, but to a landlubber the sea is always impressive and forbidding, beautiful and dangerous. The waves were reaching up and their white foam, caught by wind, was being whipped against the side of the ship.

The ship sank. Pi was 'alone and orphaned, in the middle of the Pacific, hanging on to an oar, an adult tiger in front of him, sharks beneath him and a storm raging about him'

Fear had gripped him. He was surrounded by a fierce 450 pound carnivore. The attack of the hyena on the zebra and the orangutan and their subsequent decimation is also frightening. He was seized with hunger and thirst.

Pi felt stiff, sore and exhausted, barely grateful to be still alive. He wrapped himself with the blanket and curled up on his side in such away that no part of him touched the water.

The shipwrecked animals and Pi, staying on the little lifeboat struggle against their fate to survive. Pi gives up vegetarianism and eats fish to sustain himself. The peace-loving orangutan fights greatly against the hyena. Even the seriously injured zebra fights ferociously to stay alive and demonstrates the strength of his life force. Martel shows that living creatures often do extraordinary, unexpected and heroic things to survive. When these creatures are faced with the situation of death, they will not hesitate to do shameful things. The hyena's treachery and the blind Frenchman's turn towards cannibalism are such examples.

The characters in 'Life of Pi are naturally concerned about food and water. It is ironical, though, that life surrounded by food and water offers little food and water to the inhabitants. The sea-water being salty is undrinkable and food is difficult to catch. Pi constantly struggles to land a fish or pull a turtle up over the side of the craft, so that he must consistently collect fresh drinking water using the solar stills. Pi transforms himself. He becomes a non-vegetarian because if he has to survive he must embrace a new life.

In '*Life of Pi*', the colour orange symbolizes hope and survival. Just before the scene of the sinking of the ship, Pi's daughter Usha is shown carrying an orange cat. Through this symbol, Pi is guaranteed to survive the catastrophe. The little orange cat recalls the big orange cat, Richard Parker, who helps Pi survive during his 227 days at sea. "As the Tsimtsum sinks, Chinese crewmen give Pi a lifejacket with an orange whistle, on the lifeboat, he finds an orange lifebuoy. The whistle, buoy and tiger all help Pi survive, just as Orange Juice, the orangutan provides a measure of emotional support that helps the boy maintain hope in the face of horrific tragedy. (Sparknotes)

Pi in the company of Richard Parker, a ferocious tiger on the lifeboat makes several plans to get rid of him. He makes six plans to finish him off. Push him off the lifeboat, kill him with the six morphine syringes; attack him with all available weaponry; choke him; poison him, set him on fire, electrocute him; and wage a war of attrition (vide chapter 54). Then he came to plan seven in chapter: Keep him alive.

He decides to be daunted, but not to be defeated. He decides to remember: the spirit, above all else counts. If one has the will to live; he will.

Pi also practices some rituals to keep the tiger at bay. But the greatest thing that Pi does to save himself from Richard Parker is the marking off a territory between himself and the tiger. Animals are territorial creatures. Tigers are similarly territorial. They mark their space and define its boundaries carefully, 'establishing absolute dominance over every square inch of their area. To master Richard Parker, Pi must establish his control over certain zones in the lifeboat.

He pours his urine over the tarp to designate a portion of the lifeboat as his territory, and he uses his whistle to ensure that Richard Parker stays within his assigned space. The small size of the life boat and the relatively large size of its inhabitants make for a crowded vessel. In such a confined space, the demarcation of territory ensures a relatively peaceful relationship between Pi and Richard Parker.

There is another reason for a peaceful living between Pi and Richard Parker. It was Pi, who made arrangement for Richard Parker's food and water:

> "I was the source of food and water. Richard Parker had been a zoo animal as long as he could remember, and he was used to sustenance coming to him without his lifting a paw. Time, when it rained and the whole boat became a rain catcher, he understood where the water came from. And when we were hit by a school of flying fish, there too my role was not apparent. But these events did not change the reality of things, which was that when he looked beyond the gunnel, he saw no jungle that he could hunt in and no river from which he could drink freely. Yet I brought him food and I brought him fresh water. My agency was pure and miraculous. It conferred power upon me. Prof: I remained alive day after day, week after week. Proof: he did not attack we, even when I was asleep on the tarpaulin. (*LoP*: 223).

This paragraph demonstrates peaceful co-existence between man and beast.

Pi, the castaway, finds himself in grim and exhausting opposites. The stranded person or living creature has to make adjustments if one wants to survive. One gets one's happiness where one can. One reaches a point where one is at the bottom of hell, yet one has one's arms crossed and a smile on one's face, and one feels that one is the luckiest person on earth. Pi says that life goes on because nature understands that human life should be provided for ("Why? Because at your feet you have a tiny dead fish). This interconnectedness between man and nature establishes a balance between man and nature and underscores their everlasting relationship:

> Otherwise, to be a castaway is to be caught up in grim and exhausting opposites. When it is light, the openness of the sea is blinding and frightening. When it is dark, the darkness is claustrophobic. When it is day, you are hot and wish to be cool and dream of ice cream and poor sea water an yourself. When it is night, you are cold and wish to be warm and dream of hot curries and wrap yourself in blankets. When it is hot, you are parched and wish to be wet. When it rains, you are nearly drowned and wish to be dry. When there is food, there is too much of it and you must feast. When there is none, there is truly none and you starve. When the sea is flat and motionless, you wish it would stir. When it rises up and the circle that imprisons you is broken by hills of water, you suffer the peculiarity of the high seas, suffocation in open spaces and you wish the sea would be flat again. The opposites often take place at the same moment, so that when the sun is scorching you till you are stricken down, you are also aware that it is drying the ships of fish and meat that are hanging from your lines and that it is a blessing for your solar stills. Conversely, when a rain squall is replenishing your fresh-water supplies, you also know that the humidity will affect your cured provisions and that some will probably go bad, turning pasty and green. When rough weather abates, and it becomes clear that you have survived the sky's attack and the sea's treachery, your jubilation is tempered by the rage that so much fresh water should fall directly into the sea and by the worry that it is that last rain you will ever see, that you will die of thirst before the next drops fall." (LoP: 216-217)

In fact, what appears to be opposites in nature are different facets of nature. Man has to learn to adjust himself to them to live peacefully.

Conclusion

In this paper, an attempt was made to define ecocriticism. Ecocriticism is the study of interrelationship between man and environment.

It was also shown how Yann Martel's 'The Life of Pi' lends itself to ecocritical interpretation. The novel under question demonstrates this interconnectedness in several ways. The hero of the novel owes his name to a Parisian swimming club entitled Piscine Molitor (Patel). His association with animals can be seen in that he lives in a zoo owned by his father in Pondicherry; he studies zoology at the University of Toronto for his B.S. degree.

He presents a detailed description of animal behavior, flight distances, dislike for encroachment, territorialism, and cruelty perpetrated by visitors on the zoo animals. The description of the zoo and its animals has been presented from a human angle.

There is a detailed description of the natural cycle of food chain. The narrator is shipwrecked and finds himself on a lifeboat in the company of a zebra, an orangutan, a hyena and a 450 pound ferocious tiger. Pi is forced to spend days on the vast Pacific Ocean with these animals. To his partial relief, the natural cycle of food chain operates and he notices that the hyena kills the zebra and the orangutan who both put up a heroic fight against the hyena. The hyena later on is killed by the tiger.

The novel is full of descriptions of sea life, natural phenomena (rain, storm, lightning) a carnivorous island, birds, sharks, fish and a host of other interesting natural facts and details.

There is beauty, grandeur and existence of opposites (smoothness vs roughness; sea of tranquility vs tumultuous ocean) in the ocean but Pi learns to reconcile and adjust himself in the hostile atmosphere of the sea. He also makes an ingenuous escape form a blind carnivorous French man and the carnivorous island. He also discovers that man because of his selfishness pours tons of waste on the sea-shore thus disturbs the ecological balance.

Pi learns to live with the carnivorous tiger by applying the Principle of territorialism, whistling and by controlling and mastering him through peaceful co-existence. He becomes his provider of food and water. It is shown in the novel that by keeping a balance between man and nature man can survive in the most difficult circumstances in nature.

Notes and References

1. The British Council has published a new anthology entitled "Feeling the Pressure, the Poetry and Science of Climate Change'. This anthology has been edited by Paul Munden. British leading poets have found a place in it. The present poem occurs in 'Connecting' a British Council Journal April-May 2008, p. 21, ed. By Rod Pryde.
2. *Science Reporter*, July 2008, P. 53. and *Times of India* 19th May 2008.
3. Refer to http://www.asle.umn. edu/conf/other-conf/wla/1994/ html also http://asle.umn.edu/archive/intro/sierra.html
4. "The Environmental Imagination" by Buell Lawrence 1996.
5. http://en.wikipedia.org/wiki/Food-chain
6. Science Reporter, July 2008, P. 53.

Works Cited

Atwood, Margaret. 1968. *The Animals in That Country*. Toronto: OUP.

Cohen, Michael P. 2004. Blues in the Green: Ecocriticism Under Critique. *Environmental History* 9.1: 9-36.

de Blig. H.J. 1982. *Encyclopedia Britannica*, Vol 6 and 12.

Estok, Simon C. 2001. A Report Card on Ecocriticism. *AUMLA* 96: 220-38.

Glotfelty, Cheryll and Harold Fromm. 1996. *The Ecocriticism Reader, Landmarks in Literary Ecology*. ed. Athens and London: The University of Georgia Press.

Hamilton, Joan. 2000. *Nature* 101. Sierra: 48-55.

Heise, Ursula K. 1997. *Science and Ecocriticism*. The American Book Review 18.5: 4+.

Jones, D.G.1970. *Butterfly in the Rock*. Toronto: University of Toronto Press.

Marshall. 1980. *Tom Harsh and Lonely Land*. Vancouver: UBC Press.

Martel, Yann. 2001. *Life of Pi*. New Delhi: Penguin Books India Pvt. Ltd.

Pryde, Rod. *'Connecting' May-April 2008*, New Delhi: British Council.

Rueckert, William. 1978. "Literature and Ecology: An Experiment in Ecocriticism" *Iowa Review*. 9-1 writer.

Vannucci, Martha. 1994. *Ecological Readings in the Veda*. New Delhi: D.K. Printworld.

Worster, Donald. 1993. *The Wealth of Nature: Environmental History and the Ecological Imagination*. New York: Oxford University Press.

Wynne Davis, Maarrion. 1990. *Printice Hall Guide to English Literature*. London: Printice Hall.

13

An Ecocritical Expedition into Shakespeare's *Cymbeline*

— Sumana Biswas

I

Nature has been inevitable reigning agent in the fields of science, technology, economics, history, literature, sociology and all such institutions of knowledge despite their chronological and generic heterogeneity. Human beings and Nature are complementary to each other, sharing the responsibilities of creation and its continuance. Neither he is superior nor is she inferior. But when man tends to conquer, manoeuvre and commercialize Nature exorbitantly, an acrimonious confrontation followed by separation between man and Nature hails as an aftermath. This distorted activity of man against Nature has certainly left an impact on society and literature.

Human civilization is a part of the eco-network. Therefore, it is necessary for human beings to abide by the constitution of Nature in order to maintain an ecological equipoise. To fulfill the responsibility a consistent, non-corrupt, adaptable, compassionate, patient and impartial temperament is required. Also, practicing of brotherhood and non-violence, and promoting of freedom are mandatory for peaceful co-existence. Furthermore, civilization should incorporate in it, a propensity for restoring the damage caused to the environment. All these characteristics render an individual an eco-affable entity.

Ecocriticism equally focuses upon the flamboyant and cryptic facets of sovereign Nature that stimulate the civilization integrally. Cheryll Glotfelty says that Ecocriticism deals in the interconnection between Nature and culture; it is a theoretical practice/ discourse, which acts as a bridge and negotiator between the human and the non-human. Scott Slovic describes the term "ecocriticism" as "either the study of nature writing by way of any scholarly approach or, conversely, the scrutiny of ecological implications and human-nature relationship in any literary text, even text that seem (at first glance) oblivious of the nonhuman world.[1] "Ecocriticism expounds the intertwined position of the zoo/bio-centric world, women, and the 'discriminated others in the civilization. It rejects anthropocentrism, critiques instrumental rationality and voices for eco-awareness because it believes in the fact that richness and biodiversity are valuable in themselves and humans have no right to reduce this diversity. One should care for the other life forms as a part of individual self realization rather. "Many perceptive minds have pointed out time and again that our thinking and perception have been determined by the material/ corporeal environment rather than the natural. Our constructions of our environment and our lives have become so removed from the organic unity of the poetic and the spiritual and so how could we sense and see the elemental harmony. Ecocriticism is further an attempt to reintegrate the human and the non human, to retrace the lost links between humanity and the world out there."[2]

The intended enterprise concentrates upon reading in an ecocritical dimension, Shakespeare's dramatic romance, *Cymbeline* (1610), which chiefly depicts power struggles, vacillating human relationships, and anthropocentric manipulations along with their destructive influence. Fortunately, all end well because of the strong forces of preservation. It attempts to evaluate whether a human being represents ecocide, eco-duality, eco-supportive, or eco-congeniality either lead astray or slain by ecocidal elements in the play.

II

Cymbeline's daughter, Imogen silently submits to his outrageous decision of shifting Posthumus from England to

Italy, an intentional act of separating Posthumus from her as a punishment against their audacity of marrying each other despite their unequal descent. Here she falls prey to barbarous anthropocentricity opposed to the law of Nature. Despite her strong dislike of Iachimo's licentious advancements and offending comments against Posthumus, Imogen handles the situation calmly. Thus Imogen's truthfulness, rationality and monumental patience are evinced. Imogen consents to Iachimo's feigned request of keeping a trunk full of valuable riches in her custody for one night in Act I Scene VI. Both events project her as an amalgamation of ingenuousness and adaptability. Imogen's charitable self like Nature's is taken undue advantage of by Iachimo who hides himself (instead of presents) in the trunk and treasonably trespasses into her bedchamber in Act II Scene II. She has been commercialized inordinately by the Queen who endeavours to use her as a resource for enriching and ensuring her son's materialistic sustenance. Nevertheless, Cymbeline considers Imogen his own harvest and so, desires/assumes to manipulate her. Posthumus views her as his possession, which he not only puts on stake but also decides to obliterate for pacifying his self-created psychological turmoil later. Cloten tries to seduce Imogen as he perceives her as an asset that can fetch carnal pleasure and bright prospects for him. She is no more than a trophy of conquest for him. Furthermore, her grave emotional distress leads her to the extent of commodifying her own being, "Poor I am stale, a garment out of fashion;/... I must be ripp'd: to pieces with me!".[3] (3.4.49-51) Thus her status equates that of land/ geographical region or bio/zoo-centric world often viewed by civilization as a commodity meant to be possessed and juggled in any way it wants.

Imogen's courageous act of marrying Posthumus and her tenacity of defending her husband's honour evince her explicit wifehood outliving her daughterhood though both roles are anthropocentrically designed ones intended to limit and maneuver feminine gender. She remains placid, uncomplaining and contented despite being disinherited from future enthronement by Cymbeline who transfers the right to his lost sons on being reconciled with them. Here she conforms to the conventional roles of ideal sister and daughter. Imogen's selfless service to the extent of abrogating herself towards Posthumus gets exhibited in her request to Pisanio to kill her in order to quench the murderous thirst of Posthumus: "Do thou thy master's bidding:/ ...I draw the sword myself: take it, and hit/ The innocent mansion of my love....". (3.4.63-70).

She sacrifices her whole being in performing her duty, in maintaining and preserving relationships around her under different banners for sustaining peace. Thus, she plays the role of a maintenance worker like Nature. Her opinion/ voice, rights and freedom have been ignored and suppressed like that of the bio/zoo-centric world with the same totalitarian consumerist temperament by civilization in pursuit of either corporeal benefits or perverted pleasure. Imogen's emotional onslaught by Cymbeline, Queen, Cloten, Iachimo and Posthumus parallels her with the ecosphere ploughed and undermined by unkindly civilization. She is a victim of the conservative policies dictating that women like Nature, the primary productive and indispensable supporting agent of the bio-cycle need to be maneuvered by patriarchal civilization. Thus, she manifests the intertwined subordination and agony of women and Nature. Moreover, Imogen's chaste and innocent being is maculated by Iachimo deceitfully for establishing his prejudiced argument against the chastity of women in his comment, "If you buy ladies' flesh at a million a dram, you cannot preserve it from tainting...." (1.4.131).

This patriarchal statement is illogical ecocritically because any living creature's flesh decay and decompose after expiry irrespective of species and sex in impartial Nature's custody unless they are artificially preserved. Imogen's plight reminds one of Nature's sanctity, human beings have been exploiting and defiling ever since for their egotistic and selfish gratification. Posthumus's suspicion (an anthropocentric feature) leads him to the verge of scheming Imogen's execution. Imogen, being an entity of the ecosphere has right to live and possesses an intrinsic value despite societal narcissism. Such attempts of confining, degrading and killing Imogen typify civilization's abusing of eco-ethics since the acts are inherently ecocidal. Imogen's revolt against insensitive prejudiced mercenary world to defend her honour and identity can be symbolically associated with Nature's rebellion against continuous undue depletion by civilization. All the events throughout the play evince that Imogen like bio/zoo-centric beings have always tried to adapt as required for maintaining a healthy network of life. She harbours respect for others, cherishes propensity for amiable co-existence and proves her concordance with rhythm of the environment. Moreover, she believes in destiny, a phenomenon of Nature, camouflages (a natural technique used by bio/zoo-centric entities for biological reasons) herself as a page boy and takes shelter in the forest of Wales. Thus she throws herself into the tutelage and custody of philanthropic Nature to protect and preserve herself from civilization's frigidity. Imogen's ascending impoverishment is not due to any fatal flaw in her, but due to non-acknowledgement of her efforts, and constant exploitation, abrogation, depletion, and subjugation. She is an embodiment of eco-affable ideologies and personification of magnanimous and tender façade of Nature.

Pisanio sincerely devotes himself in service of his master Posthumus and Imogen. The concept of servitude itself is a human-centred issue against law of Nature. His prime purpose is to preserve Imogen against all odds in physical absence of Posthumus. He nullifies Queen's repeated attempts to hijack him from Posthumus. Cloten's bribery, "Wilt thou serve me?.../ Give me thy hand; here's my purse." (3.5. 122, 125) and sword, "...thy silence on the instant is/Thy condemnation and thy death" (3.5. 98-99) too fail in overpowering him The Queen's wicked intentions of consuming Imogen for corporeal benefits, Cloten's harassment of Imogen for carnal gratification, Posthumus's arrogant, violent, calamitous and homicidal disposition against Imogen, Cymbeline's callousness and inefficiency in gauging Imogen's danger as a dutiful father, and Imogen's helplessness and predicament are the distressful matters which Pisanio understands with his panoptic rational vision. He tactfully handles the situation, when he is ordered by Posthumus to kill Imogen.

He assists Imogen to flee from Britain and also disobeys Posthumus's murderous orders. By doing so, he acts as an agent of Nature and desists from tainting and destroying Nature personified in Imogen. Pisanio's pragmatism and concern for Imogen gets reflected in his advising and helping Imogen in her adversities to quit Britain in male-disguise while he would send "some bloody sign" (3.4.124) to Posthumus and it will be confirmed by "You shall be miss'd at court" (3.4.125). His such actions reflect rejection of anthropocentrism and his respect/care for others as part of individual self realization. Posthumus's imposition of his lethal command upon Pisanio regarding Imogen's slain, "I expect my revenge...Pisanio must act for me/...Let thine own hands take away her life" (3.4. 22-24) shows how Pisanio's feelings have been ignored/ nullified like that of domesticated and yoked zoo-centric entities. He has undergone commodification, exploitation, degradation and subjugation in the hands of social hierarchical bureaucracy despite his honesty, dutifulness and eco-supportive disposition. He wisely follows the nonhuman world in adapting to circumstantial necessities and succeeds in contributing his part to the restoration of the environmental balance by preserving the good (Imogen). Simultaneously, he is aware such that no harm is consciously caused by him. Pisanio's ignoring of mercantile temptations, threats or discrimination practised in human empire signifies his selfless, incorrupt, patient, responsible, dauntless and unhostile temperament despite dwelling in a material/corporeal environment. His enterprise/ motivation to save Imogen from being eroded away by the synthetic circle of destruction and decay identifies him as one inclined to conserve ecological virtues and life-sustaining elements of Nature from being scourged by an anthropocentric civilization. Therefore, Pisanio gets established as an eco-congenial individual.

Belarius and his loyal services, instead of being prized, are condemned as treason by Cymbeline to the extent of his banishment from the kingdom, though all unjustness against Belarius is claimed as an accident by Cymbeline later. Thus Belarius's position equates that of bio/zoo-centric entity often used, overlooked and maltreated indifferently by the civilization despite their utility and aid to mankind. Such abrogation enrages him and as a gruesome consequence, he abducts the infant princes, the heirs to Cymbeline with the help of Eurephile, their nurse to feed his appetite of vengeance/ revenge, an eco-antagonistic phenomenon:

> Am that Belarius whom you...banish'd:/...These gentle princes,-/ I train'd up:...my breeding/...Their nurse, Eurephile,/ Whom for the theft I wedded, stole these children.. (5.5.333-341)

In the wilderness of Nature, a predator's stealing of eggs/ babies from the abode of other birds/animals for sustaining itself and its own offspring is a common inevitable biological practice, a process of maintaining equilibrium in ecosystem. But, Belarius's deliberate deprivation of the infants from biological parental care does not pertain to any bionomical contingency and is a crime against Nature. Such an act on his part exhibits his lack of foresight, diversity and ecological rationality.

However, Belarius fosters Guiderius and Arvarigus, Cymbeline's sons with utmost care amidst Nature (the forest of Wales). In a way he plays the role of a maintenance worker formerly for Cymbeline and in due course for the princes. Engulfed by his duty of rearing them up, he deliberately marginalizes himself and Eurephile from natural parenthood. It indicates abuse and denial of freedom for the natural order and repression of the instinctual life in quest for objective rationality. His economical use of the forest-wealth, his gratefulness and respect towards bio/zoo-centric world gets reflected in his attempt to convince the princes "O, this life/ is nobler.../Richer.../Prouder than rustling unpaid for silk" (3.3.21-24). Such a disposition of Belarius reveals his harmonious and cordial relationship with Nature, his dread of man-made closed space and his constant endeavour to motivate the princes away from the synthetic world. The textual excerpts simultaneously indicate how his role as a nurturer/protector gets overshadowed by authoritative conduct towards Guiderius and Arvarigus. Belarius anthropocentrically maneuvers and restricts their vigour and freedom against ecological norms.

He assertively imposes his experiences and opinion upon them, resents to provide them space and refutes their view unkindly. Ecocritically, such an act illustrates man's inherent inclination to demonstrate power and authority and non-realization of the fact that not through domination and assertion of right and possession, but only by an aesthetic distancing, one can relish the universe. Thus, his heterogeneous behaviour indicates that he oscillates between eco-sensibility and anthropocentricity. Anyhow, his Nature-intimacy enigmatically inculcates and enhances in him the spirit of rectification, which eventually make him confess his outrageous wrong action and restore the princes to Cymbeline, their biological father: "...gracious sir, here are your sons again." (5.5.347-348) Thus, ecologically he repairs the losses that he had inflicted upon the environment. In return, Belarius's emotional bruises are healed by Nature in the form of repentance, forgiveness and consolation from Cymbeline, the representative of civilization and through his reinstatement into his suitable habitat. Belarius, therefore, can be identified as an eco-dualistic character.

Guiderius and Arvarigus grow up in the forest of Wales. Their survival in a different habitat has been possible because they have learned to adapt naturally like bio/zoo-centric entities. They not only admit the inevitability of the interrelatedness/ interdependence of landscape and its inhabitants but also adore Nature's providence that has been sustaining them since their infanthood. They restrain from inordinate and remorseless exploitation of the natural resources and rather use it economically for sustenance like non-human entities.

Thus, they contribute in maintaining the ecological balance in the forest. Their rational and sympathetic disposition towards helpless, disguised Imogen who is an alien to the forest and has happened to take shelter in their cave mirrors their understanding the eco-phenomenon of cooperation and accommodation. The fact that they continuously compromise, despite their adventurous mind getting fascinated towards the glamour of civilized empire, evinces their regard for Nature. Nature has rendered them sufficiently stable, harmonious, cooperative, non-toxic, dynamic, and righteous for the conservation of a balanced human relation with the ecosystem. Their cultural values are sensitive to ecology; the influence of bio/zoo-centric world gets reflected in their manners. They embody how life outside man-made habitat thrives gloriously and temperament becomes more panoptical in contrast to those amidst constrictions of an anthropocentric domain. But, despite all their Nature-affinity, at a certain moment, Guiderius complains about the monotony of a marooned life far from civilization. Arvarigus too disapproves their savage-like life, which is being maneuvered by Belarius:

> "...Our pinching cave.../The freezing hours...We have seen nothing/ We are beastly...our cage...the prison'd bird/ And sing our bondage freely." (3.3. 37-43)

Such depression and discontentment within them is actually an aftermath of their subjugation and colonization by Belarius. Also, they have been deprived of their right to experience the kaleidoscope of human empire by Belarius in the same way as the non-human beings are marginalized from their biological habitat through domestication. Their protest is against their confounded identity and status, the denial of freedom and the suppression of their desire in Belarius's quest for objective rationality in the name of eco-affability. Their agitation is ecologically reasonable, yet it is vivid in their conversation and does not take the form of violent outburst. Thus, their patience and conflict avoiding temperament get projected here. But, Guiderius and Arviragus ultimately outrule Belarius's cautious objections and step into the material world from the wondrous and tranquil realm of Nature. In the process, they disgrace Nature's wilderness as "a cell of ignorance" (3.3.27-33) out of their avidity for corporeal progressiveness. It exteriorizes their unjust criticism of Nature's richness and biodiversity and documents their eventual turning away from Nature. Thus, degradation of eco-cordial temperament into anthropocentric attitude in human beings despite being reared in organic environment gets reflected here. Guiderius and Arviragus, therefore, personify eco-dualism.

Iachimo provokes Posthumus to enter into a wager surrounding the chastity of woman/Imogen, and plays with his sentiments for obtaining perverted pleasure. He injects into Posthumus ecocidal toxin of suspicion and envy against Imogen artfully and lethally poisons conjugal bliss of Imogen and Posthumus mindlessly in order to justify and establish his personal prejudiced opinion of women. His reprobate ambition of proving woman's frailty and avenging his insult from Posthumus in defence of Imogen's honour in Act I, Scene IV leads him to overexploit Imogen's innocence. Here, winning the wager by implementing ecocidal weapons of wickedness, fraudulence and treachery to satisfy his destructive human ego indicates civilization's demonstration of anthropocentric intellectual power through materialistic conquest. It is documented (as per Act II Scene II) in his entry into Imogen's private premises by hiding in the trunk and then stealing every minute detail provocative enough to perturb/shatter Posthumus. Ecocritically, Iachimo typifies civilization, which proves its cataclysm even in its infancy when compared to earth's geological time scales.

His statements, "If you buy ladies' flesh at a million a dram, you cannot preserve it from tainting...."(1.4.131)and "If I bring...testimony...I have enjoyed the dearest bodily part of your mistress, my.. ducats are yours...." (1.4.143-149) show how he anthropocentrically commodifies and impoverishes woman, an entity of Nature. It marks him as an uncompassionate, consumer-oriented capitalist against ecological standards. His defilement of Imogen equates civilization's abusing of Nature. Thus, dearth of organic sensitivity and spirituality gets reflected in him. But, later, Nature's mysterious driving force canes him down with sense of guilt such that he undergoes penance. In this manner Nature provides him the opportunity to rectify and reform himself. Iachimo's confession of his obnoxious deeds, "I was taught/ Of your chaste daughter..../ 'Twixt amorous and villainous./ Being thus quenched.../ To make the noble Leonatus mad, By wounding his belief.../ With tokens thus and thus" (5.5.193-203) not only revives and rehabilitates the dying relationship between Imogen and Posthumus, thus counterbalancing the lost elemental harmony but also purge his polluted soul in the same way as the considerable amount of rainwater while flowing on the land washes it of the garbage and pollutants, thus acting as an eco-cleanser. Iachimo's role, therefore, flows from the vessel of an ecocidal agent to that of a restorer.

Cloten's dependence apparently appears like that of a parasite that depends upon the support of host entity for nutrition and growth. He depends upon his mother for ideas, Cymbeline for resources and Imogen for future prospects. But, it is a bionomic requirement for a parasite while Cloten's cause is entirely a worldly one. He pursues and handles every task with an insensate and technological outlook under vicious tutelage of Queen. Cloten forcefully and wrongly intervenes into Imogen's territory to seek triumph over Imogen for demonstration of power and authority. Ecocritically, he lacks a panoptic insight of understanding the emotional distress, necessities and real self of Imogen. His material rationality views her only as an asset that will fetch him corporeal advantages. Furthermore, he desperately wants to construct his dream castle upon the ruins of Posthumus. It equates him with non-adaptable colonial consumer-oriented capitalist civilization, which undermines Nature without any concern for her replenishment or conservation to fulfil its monumental aspirations. Cloten's statement, "...the bitterness of it I now belch from my heart—that she held the very garment of Posthumus in more respect than my noble and natural person...will I ravish her: first kill him/...and I'll be merry in my revenge." (3.5.136-146) document his nesting of avarice, malice and envy, the ecologically fatal elements that direct him towards non-redeemable ruin. His recurring failure of attaining Imogen, a Nature embodiment, through temptations and threats evinces diminished existence of human beings and their anthropocentric methodologies in front of Nature's will. His mansion nesting evil intentions collapses because it is a fallacious structure in a morally sound environment constructed by Imogen around herself. Cloten's self-initiated physical encounter with Posthumus, Guiderius and Arvarigus mirrors his fostering of violence, an ecologically disasterous phenomenon. His death in the hands of Guiderius and Arvarigus in the mountainous forest of Wales near Milford

Haven shows derangement of anthropocentrism by Nature-nurtured entities in alliance with Nature. His ecological irrationality in trying to be autocratic through undue intrusion perishes him ultimately. Thus his role remains constrained to that of a destroyer; hence he is an ecocidal agent.

Posthumus's initial disposition is quite noble, calm, benevolent, dutiful, strong and dynamic. His chivalrous self has been associated with that of "an eagle" (1.1.139). His patriotism reflects his dedication towards protecting his nation from alien invasion, and encroachment. Such a natural tendency of defending one's own territory from undue intrusion equates him with the bio/zoo-centric beings. It also exhibits how Posthumus is sufficiently indebted to the environment he belongs to and pays his gratitude by being serviceable and dutiful to the same. His patience gets evinced when he silently endures his separation from Imogen in the name of deputation to Italy as per Cymbeline's orders. His overflowing love for Imogen indicates his eco-amicable temperament.

Furthermore, he counterattacks Cloten's vehemence against him and evil intentions towards Imogen with violence for defending himself and Imogen's honour. By doing so, he espouses the law of Nature which entitles all entities to exercise the methods of self-defence against the invasion of predators. His employing of Pisanio at the service of Imogen in order to shield her from all adversities during his absence illustrates his profound concern for Imogen. Despite being an entity of mechanical human habitat, Posthumus's perception and ideologies, before being deported to Italy, manifest ecological rationality. But, human views are so influenced by technological environment far from the natural, that in the man-made domain, Posthumus's silence, docility and eco-centric virtues get inordinately capitalized by the opportunistic Queen to manipulate his exile for fulfilling her purpose of furnishing and securing her son Cloten's future prospects. His state, thus, resembles that of biosphere which is ravaged for mercenary reasons by civilization.

He is forcefully isolated from Imogen as a punishment against his audacity of marrying the princess despite his lower descent. Thus, he, like bio/zoo-centric entities, undergoes anthropocentric discrimination practiced against Nature's law of equality. Moreover, Posthumus's deportation from Britain to Italy smothers him with dissonance, bewilderment and inconsistency because of his mental/idological incompatibility with a new/alien environment. Posthumus's thoughts and movements get contaminated and paralyzed consequently. Posthumus's separation from Imogen corrodes his life-sustaining energy, psychological stability and his lineament. Survival of an entity under such conditions is possible if only it is competent enough to sieve out its own limitations and deficiencies, and draw sustenance available from adverse conditions.

Posthumus is not adaptable enough to do so. His plight can be equated with those bio/zoo-centric creatures, which due to inter-habitat or intra-habitat transplantation shrink or become extinct eventually thereby heading towards an ecological chaos. Gradually, he falls prey to Iachimo's sadistic intentions under the guise of a wager. He fails to overcome his temptation of participating in the wager and unjustly stakes Imogen's fidelity and chastity for an egocentric gratification: "...let there be covenants drawn between us: my mistress...I dare you to this match." (1.4.136-141)

His thinking gets corrupted by somatic anthropocentricity and he yields to prejudices, waywardness, insensitivity, irrationality and inconsistency. Resultantly, he first commodifies Imogen and then tramples her under the burden of suspicion. His mindless and barbarous order of killing Imogen shows his deteriorating concern for Imogen and violation of the fact that murder or deliberate annihilation practised by humans is an ecological crime and that the civilization has no right to reduce any form of life either. Posthumus, the defender of Imogen, becomes her predator in the same way as civilization, the so-called protector of Nature turns into an assailer of the same. His insensible mal-actions under the influence of his prejudiced imagination simply aim at pacifying his temporal discontentment and unreasonable rage, besides contaminating the entire atmosphere. It is synonymic to the hazardous environmental pollution caused by spreading of industrial toxins innovated by man in the name of progress and necessity. Thus his eco-antagonism gets highlighted. Posthumus's shutting out of Imogen symbolizes his turning away from Nature heading towards his downfall and discontentment.

Much later, he is found repenting his rash actions and wishes to live no more for the injustice he has done to Imogen. He also participates in the battle with the Romans for the defence of his nation and not for a colonizing mission. So, though the battle and Posthumus both represent violence, blood-shed and destruction, yet Posthumus can be exempted from the allegation of toxicity since it indicates his revival of dutifulness and compunction. His ecological wisdom that got buried under the rubble of destructive and exploitative philosophies eventually re-emerge from the avariciously acidified and eroded state of his mind just like the rise of phoenix from its ashes. He is re-nourished by his tears of remorse in a miraculous manner as though driven by Nature. In a prison of Britain, (see 5.4.30-122), while he sleeps ('sleep' is a Nature's process and blessing to restore tranquility of mind and brain), a collection of spirits ascends from the infernal region and gather around him. They are Posthumus's dead ancestors, and they plead with Jupiter, the king of the gods, to take pity on their descendant and restore his fortunes. After a time, Jupiter himself arrives from the heaven, surrounded by thunder and lightning and riding on the back of an eagle:

> "He came...thunder...celestial breath...the holy eagle/
> Stoop'd...His ascension...." (5.4.114-119)

Jupiter's power is accentuated because of his affinity with the natural objects. He agrees, and Posthumus awakens, feeling miraculously refreshed, and finds beside him, a written oracle which he is unable to interpret. The oracle epitomizes mystery of Nature that often remains incomprehensible for humans.

The entire dream-like development symbolically suggests the commencement of his restoration. It is followed by Cymbeline's and Imogen's forgiveness along with his reunion with Imogen. This chain of events can be construed as an enigmatic device of Nature, to reinforce, reinstate and preserve the virtuous. So

Posthumus impersonates how eco-congeniality is lead astray by ecocidal elements followed by the revival of the eco-ethical.

Cymbeline lets himself to be manoeuvred by Queen personifying overweening materialism: "Betwixt a father by thy stepdame govern'd." (2.1.56) Cymbeline's willful submission to Queen indicates that his thinking and perception are influenced and determined by mercenary environment rather than the natural. His outrageous declining of Imogen-Posthumus marriage marks him as an eco-adversary agent dismantling Man-Nature (Purush-Prakriti) relationship. It reflects how Cymbeline unjustly defies Nature's ways in his denial of freedom and right of natural selection to Imogen and Posthumus. Thus, Cymbeline exercises anthropocentrism unduly in the name of parenthood and kingship.

His drastic action of separating Imogen from Posthumus through banishment of Posthumus from Britain proves how social prejudices and complexities make him extremely conservative, uncompassionate, non-reasonable, non-adaptable and inert towards eco-ethics. It not only manifests Cymbeline's callous/ anserine suppression of and disregard for their natural/ instinctive passion for each other, but also his lack of care for others against Nature's principium. His nullification of Imogen's thoughts and actions, and his inclination to maneuver her renders him colonial. Ecocritically, it also shows his ignorance of the natural reality that not through domination or claiming of possession but by providing space to every entity, aesthetic delight (akin to Nature/deep ecology) can be attained. Furthermore, his instant negation of Imogen from inheritance to the throne on being reconciled to his lost sons again highlights his prejudices and discriminations pertaining to gender issues against Nature's law of equality.

His accidental accusation and banishment of loyal Belarius not only shows his negligence that causes sufferings to others but also confirms human being's proneness towards being careless and practising injustice in contrast to Nature's heedfulness, impartiality and righteousness. In the process, he has mindlessly inferiorized and commodified Imogen, Posthumus and Belarius in some way or the other against bionomical code of conduct. By acting thus, he has disregarded the fact that each entity has an intrinsic value, and hence, is significant irrespective of the human viewpoint. Cymbeline remains mute against the atrocities and autocracy of Queen because of his addiction for her; his ecological wisdom remains buried under the debris of authoritative and anthropocentric philosophies at many instances.

His failure to protect Imogen from the evil designs of Iachimo, Cloten and Queen is a symbolic representation of human being's incompetence of preserving/ conserving Nature from being depleted and degraded by the expanding corporeal appetite of civilization itself. Cymbeline is largely responsible for Imogen's impoverishment. Therefore, the inferiority and meaninglessness of mechanized mannerisms of civilization in front of Nature's authentic and quality ways is ascertained once again. Nature punishes him for a certain period by separating him from Imogen, who was apparently the only heir to his throne. Nature shuffles everything to create turmoil in Cymbeline's life and ultimately makes him detect, realize and repent his actions which were non-amicable to eco-ideologies. Destiny (Nature) provides him a scope of rectifying himself, which he does through his soft and forgiving approach towards others, freeing the Roman prisoners and accepting the reunion of Posthumus and Imogen. Thus, Cymbeline's submission to the ways of Nature, his repentance and self-realization transforms him from somatic entity to a relatively more ecologically apprehensible one under Nature's forces of conservation.

Queen's commercial temperament and overexploitation of Cymbline for her son Cloten's mercenary prospects is vividly reflected thus, " First, she confess'd she never lov'd you;/ ....Married your royalty...." (5.5.37-39)

> ... she had
> For you a mortal mineral...
> ...she had fitted you with her craft, to work
> Her son into the adoption of the crown:
> ...shameless-desperate...despite ...
> her purposes...repented.... (5.5.49-60)

In order to make Cloten a natural successor to the throne, she desperately endeavours to yoke Imogen with Cloten with the anthropocentric instrument of marriage. Ecocritically, such an act indicates not only her commodification of Imogen but also her ignoring and smashing of Imogen's perceptions and emotions remorselessly for selfish cause. Thus, denial of an individual's natural freedom and rights apart from suppression/ manipulation of one's natural instincts get reflected in her behaviour against eco-ethics. She drives Cymbeline to separate Posthumus from Imogen and ravages their peace and happiness in her materialistic pursuit against law of Nature. Her undue interference and imposition deny space to every entity. She personifies consumer-oriented capitalist culture domination, power and authority. She ambitiously tries to circumvent Pisanio (an eco-congenial agent, whom Queen views as an obstacle/fort on her way to attain Imogen) through colonization. Consequently, with a motif of killing Pisanio, Queen orders Cornelius to prepare her a deadly poison, which she claims as her requirement for scientific purposes, for application upon small animals and the like.

> He hath a drug of mine; I pray his absence
> Proceed by swallowing that....
> ... But for her ... gone she is.
> To death or to dishonour...
> Can make good use of either...
> I have the placing of the British Crown. (3.5.58-66)

It reflects her non-realization of the fact that both human and non-human forms of life have certain substantial value in the universe, and none has the right to decimate or disregard/ disrespect this ecological reality. Moreover her homicidal intention itself is an ecological abuse and depravity. Queen's indifference regarding both conservation of the environment and replenishment of the damage she causes marks her as an eco-antagonist. She lacks care, understanding and respect for others. Queen's activities project her non-adaptability, egocentricity, treachery, infidelity, artifices, obstinacy and envy, the ecologically destructive elements. To the contrary, she can be equaled with Nature on the basis of her reproductive role that renders her an elemental sustainer of her son Cloten.

When she fails, she kills herself; thus we come across her sacrifice on the altar of motherhood which totally eclipses and dissolves her other identity and functions. But, her over-zealous maternalism is so overpowered by materialism, that it turns her away from Nature and eco-ideologies. For instance, it is a characteristic feature of parents in bi/zoo-centric world to hunt for the fundamental sustenance of their offspring, which is a bionomical necessity. But in Queen-Cloten case, no such biological contingency/necessity prevails and as for Queen it is a sheer anthropocentric enterprise. Furthermore, she disguises herself with feigned goodness to predate against ecological considerations, and, thereby misemploys the camouflaging technique employed by bio/zoo-centric entities for biological purposes. She personifies degeneration, death and frigidity born of anthropocentric confinement.

She is punished by Nature through the death of Cloten in the forest of Wales. Once again, the triviality of civilization's mercenary propositions in front of Nature's will and authority get evinced in Queen's defeat. "The evils she hatch'd were not effected; so, Despairing, died". (5.5.60).The ruin she brings upon her is too total to be rehabilitated. But, finally, she contributes through her honest confession of her evil intentions and deeds in her death-bed, (perhaps for salvation), towards building bridges of restoration and acts as an agent of Nature before perishing and being assimilated into her philanthropic realms. Thus, Queen is actually an ecocidal character who ultimately gets metamorphosed into an eco-ethical being though for very short instance.

III

In Cymbeline the early episodes illustrate deterioration, decay and death of winter followed by the conceding and reconciling events manifesting the festivity of spring and summer. Queen represents anthropocentrism against eco-amicability. Iachimo personifies eco-antagonism, but later his confession contributes to restoration and evinces his Nature-driven transformation. Imogen personifies eco-congeniality despite her growth in a materialistic atmosphere. Guiderius and Arviragus are eco-dualistic entities, yet they show considerable degree of eco-compatibility. Belarius plays the roles of nurturer, preserver, exploiter and restorer; he is a fusion of eco-amicability and eco-antagonism. Pisanio epitomizes preservation and righteousness, hence ecologically constructive. Posthumus emblematizes eco-dualism to a certain extent, but again, it cannot be denied that his ecological wisdom is smothered by narrow-minded and destructive philosophies bred in a synthetic environment. Cymbeline is anthropocentric and eco-dualistic by temperament. All human beings have Nature deeply embedded in them. But in some, she gets shrouded by materialism and consumerism while in others she registers her presence clearly. All eco-dualistic and ecocidal characters in Cymbeline illustrate, how often, the course of human events and activities become so removed from the organic/natural world, that sometimes they either remain far from being regenerated or reintegrate their value systems and retrieve their sense of balance and harmony through a recognition of difference and realizing of the numerous/infinite.

Cymbeline gives voice to a host of natural forms: forest, fragrant flowers, birds, wind, wild life and caves. The ecological/ Nature has been privileged over the anthropocentric in the play. In Cymbeline, the palace, a man-made enclosure proves its inferiority/triviality thus:

> It fails to provide contentment to Queen and Cloten though the environment and their desires pertain to corporeality; it causes and aggravates frustration and deterioration in life and character of Posthumus; it pushes Queen, Cloten and Cymbeline into an inferno of anthropocentrism; it fails to protect Imogen from distress and to conserve her marriage from turmoil; it also fails to render happiness to its occupants. Furthermore, this man-made site (palace) nests conspiracy; the concrete walls cannot stop Cloten from going to his end/doom designed by Nature; simultaneously, they are not strong enough to imprison or barricade eco-affability. Thus human conception of comfortable and safe place is reversed here.

The dread of open space drives humans construct closed space which turns out to be rather a baleful region concealing ruin and burying the organic and spiritual sapience under the debris of corporeal doctrines. In contrast, life amidst Nature's tranquil realms outside human habitat thrives with vigour and instructs to be ecologically rational and adaptable. Here, the forest of Wales serves as a warm shelter providing peace and relief to those who escape from or happen to be out of the dismal glacial premises of Cymbeline's palace. It also emblematizes/signifies one's return to Nature ultimately suggesting mockery and defeat of anthropocentricity. Moreover, plenty of Nature-images have been anthropocentrically juxtaposed with both virtues and vices apart from actions, movements, physical and psychological features of civilization. Such a manipulative process witness frequent distortion of the bio/zoo-centric world in human being's paralleling it with the abhorrent images of the civilized and mercenary world too.

Posthumus has been associated with a princely eagle, Cloten to a crowing cock, Britons as crows. Arviragus associates the night to an owl and the morn to the lark. Imogen's beauty is described in terms of Nature:

> "She is fair and fresh and fragrant as the lily, pale as primrose, her veins are the tint of the azured harebell, her mole spotted as the centre of the cowslip, and her breath sweet as is the eglantine."[4]

Iachimo at first sight compares her to the fabulous phoenix "Arabian bird". Unsuspectingly, he foretells her future. She rises from a death-like sleep in the forest. Posthumus addresses her and himself respectively as, "Hang there like fruit, my soul,/ Till the tree die!" (5.5.263) Belarius alludes to the characters of Guiderius and Arviragus as "zephyrs blowing below the violet", "Their royal blood enchafed, as the rudest wind...." (4.2.173) Thus, the flamboyant and cryptic facets of sovereign Nature stimulating the civilization integrally are highlighted.

IV

The ecocritical reading of Cymbeline reveals that man's illusion of supremacy and indispensability, and his obscure/ prejudiced mannerisms have been eclipsed and dismantled as trivial phenomena by Nature's inexplicable diverse hues and omnipotence. For example, Cloten's demolition in the forest by Nature-nurtured Guiderius and Arvarigus, Queen's dismantlement of abusive intentions and corporeal ambitions indicate failure and death of civilization's mercenary projects before orphic Nature. It also confirms how Nature enigmatically maneuvers the destiny of civilization. The ecocritical analysis of Cymbeline leads us to infer that each human character plays a definite set of roles, either eco-affable, or ecocidal or a blend of both depending upon the intensity of their eco-awareness. The plays have highlighted human beings as elements of the grand set of ecosystem/Nature playing different roles like Nature and also against the ecological norms. Many a times their status remains enlaced with the bio/zoo-centric world in terms of impoverishment, exploitation, marginalization, abrogation, annihilation, inferiorisation and subjugation. Further, the play has presented anti-natural anthropocentrism against the biological world's equity and impartialit.

Ecocritically, Cymbeline has focused upon horrors and menace of civilization that have been highly instrumental in harnessing and disrupting an environment, thereby heading towards an environmental imbalance. Cymbeline has projected autocratic civilization's paucity of interest in restoring the damage caused by it in the process. Nevertheless, it has witnessed (though seldom) eco-sentience too evident in Belarius's realization of necessity to preserve the environment one gets reared in and to conserve generous Nature besides Nature-affable demeanour of Pisanio and Imogen. The gruesome aftermath of ecocidal practices and halcyon/favourable outcome of eco-friendly actions get evinced in the play. The transformation of ecocidal beings into eco-friendly entities under the forces of Nature has been also experienced. They suggest that in an ideal situation, human beings must learn to adopt the ways of Nature and their cultural values should be eco-supportive. It can be ecologically interpreted from Cymbeline not only that Nature has joy and brightness, which happen to be an integral element that flourish and revive in due course making possible the rectifications and happy restorations but also that human beings must refrain from absolute materialism and consumerism in order to avoid chaos and disorder that would direct them to catastrophic disconnection from Nature, causing their total demolition..

Notes and References

1. Estok, Simon. C. "A report Card on Ecocriticism". AUMLA: The Journal of Australian Universities Language and Literature Association 96. Nov. 2001: 220.
2. Dr. Murali Sivaramakrishnan, Green Voices: Some Aspects Of Ecological Criticism. 31 August. 2007 Countercurrents.org.
3. "CYMBELINE", Complete Works of William Shakespeare, (New Delhi : Oxford & IBH Publishing Co. Pvt. Ltd., 1980) 1197-1239. 'All textual quotations are from this edition.'
4. C. Spurgeon, Shakespeare's Imagery, (Cambridge: Press Syndicate of the University of Cambridge, 1952) 294.

Works Cited

Berman, Tzeporah. "The Rape of Mother Nature?: Women in the Language of Environmental Discourse". 1992. *Trumpeter* ISSN: 0832-6193.

Bookchin, Murray. The Philosophy of Social Ecology. Jaipur: Rawat Publications, 1996.

Branch, Michael P. et al. (eds.). 1998. *Reading the Earth: New Directions In the Study of Literature and Environment.* Moscow, IO: University of Idaho Press, 1998.

Drew, David. *Man-Environment Process.* London: George Allen and Unwin Publishers Ltd., 1983.

Hardesty, Donald. L. *Ecological Anthropology*. New York: John Wiley and Sons, 1977.

Mies, Maria and Vandana Shiva. *Ecofeminism.* London and New Jersey: Zed Books, 1993.

Waren, Karen. J. *Environmental Philosophy: From Animal Rights to Radical Ecology.* Ed. By Michael E. Zimmerman, et al. NJ: Prentice-Hall 1993, pp 253-267.

14

To Know 'Place': Proposed Curriculum that Integrates Literature and Science, Imagination and Reductionism

— Wanda Baxter

The chapter I propose considers the introduction of an undergraduate course that views 'the environment' from different perspectives of observation and interpretation. The purpose of the course is to introduce students to interdisciplinary expressions of environmental observation, in order to gain appreciation of the interdependence of creative and scientific ways of knowing, especially as they pertain to environmental sustainability. The subject matter of the course draws from place-oriented literature, poetry, film, and science-based literature.

I believe the separation of literature and the humanities from the sciences—in theory and in practice—is part of what limits our ability to respond to environmental problems. Without being able to observe and/or tell the story of a place and its people, science attempts to respond to environmental degradation via linear, detail-oriented means. What we need now is education that teaches the whole of a place. Our world is complex, and our teaching and learning curriculum needs to step up to the challenge of teaching complex ideas.

I propose this course as an introduction to the idea of 'applied literature to environmental issues'. It is ecocritical at the core, but its intent is practical. The role of the imaginative thinker and writer needs to be reclaimed as an essential part of understanding and responding to our world.

This course proposes that students of environmental sciences and planning—as well as the humanities—need to learn the value of story and imagination to reinvigorate holistic ways of thinking, and originate sustainable ways of doing.

To Know Place: Preamble

The quality of work that environmental professionals do depends on having a comprehensive understanding of how places function. Entwined amongst the complex mix of natural systems is human society—a realm with its own complexities, rules, and systemic interrelationships. With a background in environmental planning and policy, and direct experience working as an environmental professional; it has been my experience that many professionals who are in charge of managing land and land development desperately lack this kind of necessary, comprehensive understanding of human-nature ecosystem relationships or; to put it in literary terms: *they are unable to ascertain or recount the story or holistic system of the place.* The following chapter introduces a course design that specifically addresses this lack, via the practical application of literature to the needs of environmental practice.

The proposed course addresses environmental issues from different perspectives of observation and interpretation. The purpose of the course is to introduce students to interdisciplinary expressions of environmental observation, in order to gain an appreciation of the interdependence of creative and scientific ways of knowing, especially as they pertain to environmental sustainability.

Environmental observation, for our purposes, is viewed as a continuum, with descriptive, subjective expression at one pole; and literal, objective expressions of environmental elements at the other.

In the same way that we have become alienated from nature, we have also forgotten that we are nature. The course examines, also, the ways that humans have responded to nature according to what we believe and think it to be. Essentially, "To Know Place" is about learning to integrate different ways of thinking and knowing in order to develop lateral thinking skills, and to push the brain to reach beyond prior learning.

Finally, the course considers the past (changing conceptions of nature over time); it examines the present through the filters of sense of place, and science of place (i.e.'felt' experience, and the assessment of parts detached from the whole), and it explores the potential for integrating different ways of knowing 'place' as a way to frame—and respond to—the complex issues that characterize environmental sustainability.

Applying 'Ecocriticism' to Teaching Sustainability

The organized study of literature and the environment known as 'ecocriticism' exists primarily as a theoretical and self-reflective field of limited reach. The typical approach of literature and the environment studies is the consideration of artistic/personal representations of the environment, which has its place, but the practice has yet to recognize its own broader potential. The field of ecocriticism has yet to determine clear, effectual goals; it has not yet recognized where and how it can have real, and 'real-world' impact.

The development of progressive curriculum is an obvious place to start, and syllabi collected on the website of the Association for the Study of Literature and the Environment[1] reveals a proliferation of interest in examining the representation of nature and the environment in literature, poetry, and film. However, much of the curricula being developed is not substantially different from what we've always done (i.e. the analysis of texts for themes, ideas, theoretical underpinnings), only with a tighter focus on the role of environment-human interactions and the depiction of nature in writing. This work is not without merit, but it simply doesn't extend outside its own disciplinary focus, thereby diminishing its impact and relevance. Dana Phillips, author of *The Truth of Ecology: Nature, Culture, and Literature in America*, describes the direction he sees ecocriticism needs to go:

> In order to adapt itself to the vagaries of nature-culture, ecocriticism needs to be more willing to hybridize than it

> has been: it needs to have a heart and a brain as well as arms and legs ... A hybrid blend of theoretical and philosophical insight, awareness of scientific method, and a thorough acquaintance with the facts (who knows what they will turn out to be?) is necessary if we want to address nature-culture in tandem and as a singular phenomenon. (Phillips, 39)

The study of literature has always considered themes and reflected on the role of place (or 'setting'), and how writing reveals these themes; but the stakes now are higher and our role as literary scholars can no longer be as self-indulgent as it has in the past. Some will bristle at this characterization, to be sure, but if we are not teaching students how to apply their understanding of the value of place and the essential role of storytelling and writing as a vehicle for a deeper comprehension of the world around us, we are shirking a responsibility to share with the wider academic and cultural community perceptions of nature and society that are inherent to literary studies.

English scholars develop a comprehensive perception of place, of character, and the interrelations between people and the places they inhabit. While this is obvious to those of us who are immersed in literary studies, it is not obvious that other fields lack this holistic sense of place. The study of poetry, film, drama and literature involves the incremental development of a set of skills difficult to appreciate from within and without English departments. Being an English major who has gone on to study interdisciplinary Environmental Design (comprising environmental planning, law, impact assessment, ecosystem management, etc.), it is easier to appreciate the valuable role that literary studies can play in teaching sustainability and dealing with environmental problems. Without conscious intent, literary studies teaches how to 'know place', and how to think three-dimensionally about place.

The Value of Teaching How to 'Know' Place

It is a hard thing to imagine that environmental professionals will be able to respond effectively to environmental conditions, without being able to fully assess the system within which they are working. Regardless of the issue, or specialty, or profession, environmental problems are systemic: they are webs within webs of social, economic, ecological, political, hierarchical, legal, and historical conditions. The planner, or engineer, community developer, biologist, public servant, wildlife manager, or whomever enters a new scenario where they must analyze environmental conditions (whether degraded or not), and they must be able to project how their response will affect the dynamic web they are dealing with.

It is not enough to train environmental specialists. We need also to be training specialists who understand their role in the collective response. We need to be training practitioners capable of specific skills, while also being able to think laterally. We need to be teaching systems theory to every student entering environmental fields, not just those in management positions.

"To Know Place" is designed to respond to these issues, and from a perspective that literary studies have something valuable to offer. The comprehensive approach that is missing from teaching approaches in the environmental fields is inherent in literature. The critical ability that is missing is the ability to both read the complexities of a place, and the inter-reactions therein. Students of English and other humanities faculties learn characters, setting, theme, political and social scenarios, and the culture of the place: we learn how people survive and what they do to survive. In sum: by learning how to analyze and explain literary texts, we are essentially learning how to read a dynamic human system. Further, the poet/philosopher has always been interested in understanding the world, its workings, and our place in it. His/her job is the job of close and careful observation: the goal is to tease out details, behaviours, and patterns pertaining to what he/she encounters, toward developing an idea or theory or story that explains the phenomena under study. This, too, is the job of the scientist. Though their methods and objectives are different, the core of the work in both arenas stems from a similar goal: *to observe in order to know.*

The Course Overview: Teaching Environmental Knowledge from Concept to Practise

This course design began with an idea: to reveal the connections between environmental sciences and the humanities. With an interdisciplinary background that bridges the humanities and environmental design, my intent was to construct a course that specifically revealed the commonalities of these seemingly disparate fields. My intent is to share insight that is obvious to someone with a cross-disciplinary education; that being the similar approaches applied by different specialties to the same subject matter. At first glance, literary studies and applied environmental sciences and planning seem world's apart. The methods and approaches used are vastly different; where one deals with generalities, emotions and human behaviour; the other deals with specifics, is intentionally detached emotionally, and is concerned with the behaviour of (primarily) non-humans. Where one deals with psychology, ethics, and imagination; the other deals with physiology, ecology, data, and proofs. The differences between the disciplines are substantial. They represent the distinctiveness of the right and left brain. But the commonalities, once brought into focus, are significant, and they are the necessary bridge between the humanities and science that is required for teaching and implementing sustainability. This necessary bridge hinges on the common activity that is the essential activity of these distinct disciplines: observation.

The course "To Know Place" is deliberately designed as a flexible model. While the concepts and main sections of the course are intended to remain fixed as the structure enhances learning, the course material within each section can be adapted, via the use of local 'place' examples; individualized teaching approaches; concrete learning examples, and other elements. The course will be most effective if it incorporates personally-known places, regional literatures, local ecology and environmental issues. There should also be a hands-on, experiential learning element to the course. The combination of broad theoretical principals (i.e. sense of place, science of place), and tangible, detail-specific examples of these principles work together to provide an integrated; purpose-driven learning experiences.

'To Know Place': is an exploration of place, observation, and the development of environmental thought as expressed in contemporary literature, poetry, and science-based narrative. The course is designed for students at the undergraduate-level; primarily those enrolled in the humanities and environmental fields. Students of the course can be expected to have the following characteristics:

i) The students who enrol in the course will feel they know more than they likely do in—reality—about environmental thought and how it has developed over time,
ii) Learners in the course will care about environmental issues, and are interested in working to repair environmental degradation,
iii) Most students will be undergraduates of English and Environmental Studies, but the intent is to have a wide range of disciplines invited and included in the mix (increased diversity is anticipated as the course proves relevance),
iv) Expectations will be high that the course is relevant, current, and has *practical utility*.

With an idea of who the learners of the course will be, it was then important to establish the learning objectives for the course (Fink, 65). Because the course aims to bridge the distinct fields of literature and the environment, it is grounded in interdisciplinary theory and teaching methods for interdisciplinary learning (refs here). Some of the learning objectives are designed to encourage interdisciplinaryity, others are designed to teach critical thinking, systems theory, environmental ethics and history, and literary analysis. The teaching objectives of the course are listed as follows:

a) Develop an appreciation and respect for both the liberal arts and sciences with respect to environmental knowledge and problem-solving,

b) Develop an enhanced perspective of the intangible values of 'place',

c) Gain appreciation of 'observation' as a common function of both creative and scientific discovery,

d) Have working knowledge of the evolution of environmental thought and its key thinkers,

e) Be capable of reading critically, with attention to: context (historical & environmental), identifying and understanding main points, learning new terms, summarizing key ideas, and determining relevance,

f) Be capable of critiquing literature, film, science narrative, etc., for the differing ways the story of a place is told and communicated,

g) Understand 'lateral thinking', and display an ability to use this skill,

h) Become better acquainted with the idea of 'place', and how it relates to environmental sustainability (as a concept, a system, and both home and habitat) and,

i) Engage personally with (and perhaps develop empathy for) nature.

Using this information about learners, teaching objectives, and goals for the course, a concept map was developed to graphically illustrate how the course would work as a whole.

The Concept Map for the Course

> Mary Beaudry has written about the value of a good graphic image for communicating the organization and structure of the subject of a course. She argues that teachers should shift their attention from the question of "How much content is enough?" to "How should my course content be organized?" (Fink, 148)

The concept map for "To Know Place" is a key element for communicating the course design and its content. As stated previously, the bridge between the humanities and the sciences is the common activity of observation. The Concept map for "To Know Place" follows:

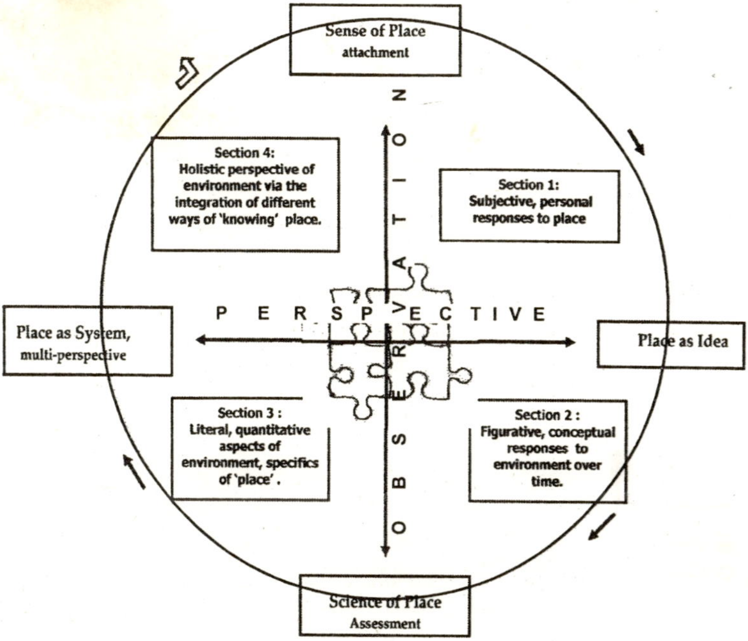

Concept map for "To Know Place"
***Applied Literature (Environment)* 3405, Wanda Baxter, 2008**

The Course Concept Map: Section Descriptions

Section 1: A Sense of Place

The concept map for the course is a useful visual representation that encapsulates the key elements of the course, and the areas within the structures where adaptation is possible and encouraged. The central subject, place, is considered throughout the course from different points of view. Beginning with 'sense of place', students are first introduced to the idea of attachment and personal experience of place.

As stated previously, the bridge between the humanities and the sciences is the common activity of observation. In this section, observation is a subjective exercise. A Sense of Place is a primary topic of poetry and literature and other creative forms. By introducing students in this section of the course to subjective writings about places, and also requiring creative personal responses to their surroundings, students begin the course from an awareness that observation can be highly personal.

This section lends well to using poetry and 1st-person narrative. Giving the students the opportunity to write about place memory and personal experiences of place is also recommended here. By initially setting up an understanding of place as an emotional, sensory and subjective experience, the course is grounded in the personal.

Place as Idea, Section 2

The second section of the course picks up on the personal-attachment elements of place and focuses more on contextuality. An 'idea of place' is distinguished from sense of place as a shift from the personally 'felt' to theoretical understanding. Seeing place in context; as a changing idea over time; an idea tied to belief systems and societal norms, the idea of place is explored through theory and ethics of human-environment interactions over time. This will involve a broad overview of how our conceptions of the world and our place in it has drastically changed over the last few centuries. Course material in the section will be chosen specifically to represent the different ideas expressed. Poetry is particularly useful for illustrating how perceptions of the environment have shifted.

Place as Idea is introduced to move the focus from observation to perspective (the second arm of the concept map). How we perceive things is filtered through our beliefs and theoretical underpinnings. The section 'Place as Idea' takes a step back from the subjective to the conceptual: the history of environmental thought is examined here (again through literary works, and delineated usefully via timetables). Observation can be viewed in this section as heavily determined by the perspective of the viewer. The main objective of this section is to develop an understanding that both the subjective and figurative are inherent to knowing place, and assessing environments.

Establishing a historical perspective of place (again, there is a role for applied poetry and literature here) and revealing how this perspective has changed over time, offers students a more layered perspective of place. This section is also designed to examine how ideas and beliefs inform our perceptions and interpretations of the environment over time. Where religion had once been fundamental to how we—as a society—perceived the natural world, the 'idea' of God as maker of all things has lost its dominance over the last two centuries. Though religion remains a primary source of knowing for a segment of the population, the rise of science and the search for tangible 'truth' has relegated belief and the nature of divine conception of nature primarily to churches and other religious houses. Poetry serves as an especially useful tool to illustrate the change of beliefs and conceptions of the environment and our relationship to it over time.

Utilizing the condensed form of poetry, we can elicit representative examples that express the personal, societal, practical, and theoretical viewpoints of an era in a single page. It is an example of applying literature to learning needs. Writing has always been concerned with the natural world. At this point in our history we need to reclaim the wisdom of perception and contemplation held in the pages of books, and realize that our teaching and learning, now, is an extension of what artists and scholars and prophets have intuited all along.

The main trends of environmental thought over time are also presented in this section. The course considered the movement from the beginnings of environmental ethics and imagination through Aldo Leopold's The Land Ethic, to deep ecology to legal entrenchments, to recent attention to sustainability, wise use, and the field of ecological criticism.

By the end of the second section, students of "To Know Place" will have gained comprehension of what 'sense of place' is, and will also be aware of the ideas and concepts that underlie sustainable behaviours and practices. They will be able to illustrate different conceptual perspectives of nature, and have a grounding in the history and development of environmental thought (and thinking) over time. The first two sections describe necessary aspects of learning the story of a place: a sense of place is an expression of 'felt' experience; while the idea of place expresses how places are conceived of and responded to. It is important that by the end of this section, students have an understanding that the idea of place has altered drastically by the time the rise of science has become the primary cultural 'lens'. To the degree that other perspectives of place have been diminished, science is more often relied upon for decision-making than our other ways of knowing.

The Science of Place:[2] *Section 3*

The third section of the course shifts focus from the personal/ conceptual pole of observation to the direct consideration of environment as object of study. The term 'science of place' means what it sounds like. A sense of place relates to general sensory experiences of an environment or region, whereas Science of place is the counterpoint and relates to specific, tangible aspects of an environment or location. The focus of this section is on hard science as a basis for understanding what a place is. Students of the course will learn the difference between analytical, reductionist observation, and the counter-puntal relationship of observing for scientific purposes—an exacting method; and observing via human-centered perspective.

A 'Science of Place' is a narrower, tighter lens that the two explored in the prior sections, and it is based on real, tangible, and this section of the course are based in concrete experiences, and involve a kind of place mapping where the environment is known through details. Students will be required to record and examine specific aspects of place (or the environment), and present their data in both analytical form, and creative visualization. The use of literal maps and field expertise is an essential part of this section, ensuring that students shift their thinking from the personal/theoretical to the practical/ applicable. This shift in thinking and/or observational stance is an essential objective of the course.

Returning to the purpose of the course: to prepare students for work in the environmental filled by learning what it is to 'know place', and the systems thinking this requires, it is key that students are made aware, throughout the course, that perceptions of place and 'knowing place' are substantially different. It is more likely that someone thinks they know a place than they really do—and for practitioners of environmental science, engineering, and other fields—knowing place must mean more than knowing the bare specifics of the science of place. For environmental activists, community developers, planners, and other fields; it is equally important that subjective and theoretical perspectives of place are balanced with hard data and collaboration with science-based place managers.

The final section of the course is designed to encapsulate the cumulative knowledge of the previous three sections, and requires that students consider place in three dimensions, via different ways of knowing. It is by no means an easy objective, but because the course builds through increments and distinct sections, systems thinking is approached gradually, through a series of stepped learning assignments, interdisciplinary learning is effectively approached through sequential learning (ref. here), and supports teaching for sustainability which demands a 'whole system of inquiry'. By the time students in the course have completed the first three sections, they will be prepared to assess problems of place and environment from a systems perspective; an awareness of the whole and its parts. This awareness lends itself directly to problem-solving that consider all the angles before trying to propose solutions. Systems thinking is appropriate to apply to environmental problems for an obvious reason: the environment is a system (or system of systems). Only through considering the systemic aspects of place, via a systemic way of thinking (i.e. multiple-perspective) will we arrive at suitably complex responses.

Place as System: Section 4

The fourth section of the course is the culmination of the integrated learning processes encountered in the beginning sections. Students will finally be required to assimilate their prior learning in this section. The mental mapping and place consciousness developed in the 'sense of place' section is used as context to acquire 'ideas of place' comprehension (here using context-mapping as a key exercise), while a 'science of place' using literal mapping of place and place specifics to gain a spatial'/physical awareness of place. In culmination, these perspectives expand in a similar way that differs from photography: from an observation expands single viewpoint to a moving, sensory, multi-perspective image. Essentially, systems thinking demands a high degree of both critical and lateral thinking. Students of the course are tasked with continual questioning, reflecting, and re-questioning. They are tasked with linking ideas, finding connections, and learning to think across disciplinary lines.

Developing systems thinking demands much of the students. They are tasked with challenging their own natural patterns of thinking, they also need to become accustomed to thinking across disciplinary boundaries.

The main exercise of a 'system of place' is to write a holistic assessment of a place, with specific attention to a particular systemic issue. The intent of the exercise is not to do an exhaustive review of a place but to be able to learn what they need to be able to write and/or document intelligently the sense of, idea of, and science of place, and to then consider the different elements as a coherent expression of place.

To Know Place: Summary

This course is proposed here as an example of how the literature part of the field of 'environment and literature'

Notes

1. www.asle.org
2. As far as I am aware, the term "Science of Place" is unique to this course concept, and to this article – its first publication. It is a useful term in this context, and I hope is a particular contribution of this work.

References

Beer, Gillian. *Open Fields: Science in Cultural Encounter*. Clarendon Press. Oxford, 1996.

Buell, Lawrence. *Writing for an Endangered World*. 2001.

Irvine, Colin. "A Novel Understanding of Ecology". *Academic Exchange Quarterly*, 7.4 (Winter 2003).

Newell, William H. "Designing Interdisciplinary Courses". *New Directions for Teaching and Learning*. No. 58, Summer 1994. Jossey-Bass Publishers.

Klein, J. T., & Newell, W. H. "Advancing Interdisciplinary Studies". In Jerry G. Gaff, James L. Ratcliff and Associates (Eds.). 1997. *Handbook of the Undergraduate Curriculum* (pp. 393-415). San Francisco.

Nikitina, Svetlana. "Three Strategies for Interdisciplinary Teaching: Contextualizing, Conceptualizing, and Problem-Solving". Interdisciplinary Studies Project. Project Zero, Harvard Graduate School of Education. 2002. *http://www.pz.harvard.edu/interdisciplinary/pdf/SvetlanaStrategies.pdf*

Phillips, Dana. *The Truth of Ecology: Nature, Culture, and Literature in America*. Oxford. 2003.

Popcock, Couglas C. D. *Humanistic geography and literature: essays on the experience of place*. London: Croom Helm; Totowa, N.J. : Barnes & Noble, 1981.

Repko, Allen F. "Interdisciplinary Curriculum Design." Mar 22, 2007. *Academic Exchange Quarterly*. *http://www.thefreelibrary.com/Interdisciplinary+curriculum+design.-a0165912657*

Second Nature. "Sustainability Curriculum Framework". www.second nature.org.

Sherlock, John. "Pedagogical Tools to Develop Critical Thinking". In: *Academic Exchange Quarterly*, Dec. 22, 2005.

15

Representation of Nature in Ancient Indian Literature: Some Reflections

— Ujjwal Jana

In the contemporary literary parlance, the term 'ecocriticism' has gained wide currency in view of the growing global concern for the degradation of ecological landscape, climate and the environmental bodies. What do we mean by the term ecocriticism? It's one of the ways in which the literary study builds up the close architectural bond between the human world and the ecological world. The literary manifestation of the physical world, indeed, grows out of the nightmarish awareness that the physical environment which is indispensible for sustaining human life on earth is almost on the verge of extinction owing to the indiscriminate use of science and technology and the mindless human exploitation of natural world. This grave environmental crisis is, no doubt, posing a threat to human existence and the need for creating a sustainable and balanced relationship between man and the natural world is one of the most urgent social and environmental issues being felt by men all across the globe in this millennium. This consciousness also paves the ground prepared for the writers to pen down the serious preoccupations involving ecology and the human culture into words. Articulating the vital intellectual part which scholars can play in their engagement with environment, noted historian Donald Worster says:

> We are facing a global crisis today, not because of how ecosystems function but rather because of how our ethical

> systems function. Getting through the crisis requires understanding our impact on nature as precisely as possible, but even more, it requires understanding those ethical systems and using that understanding to reform them. Historians, along with literary scholars, anthropologists, and philosophers, cannot do the reforming, of course, but they can help with the understanding (27).

It is tempting to explore at this juncture, how the ancient Indian literature addressed the environmental issues, as faced by people as a subject of mounting significance, many centuries back and how the nature articulated its voice in the Vedic and ancient Indian texts.

My paper concentrates on the basic premise that in Indian world view the awareness of the harmonious bond of the human nature with the physical environment was already acknowledged and a survey of the classical literature speaks not merely of the poets' profound preoccupation with, but touching sensibility to the entire gamut of biosphere-bodies. In Indian philosophical thought, the organic relation of humans to the biological nature has always been emphasized. That nature is an inseparable part of human and cultural discourse has been articulated in the *Vedas*. " ...everything that exists is an emanation from and is, in essence, Brahman (absolute) leads the Indian philosophers to use the language of ' Man is Nature', 'Man in Nature' and not 'Man against Nature'. Since it the same Divine that is manifested in Man and Nature, the Indians also avoid all issues of domination and subordination in ecological ethics" (Rama Pappu 299). This unified view of man and nature is the fundamental truth and essence of what we call Indian World view and it would not be an exaggeration to consider this ideology as a great component of Hindu religion. The modern western world view of what Mishra calls 'the Greco-Roman world-view' (58) marks a clear break with the classical Indian point of view in the sense that the former looks upon nature as the other, making it subordinate to the dominant discourse of man. The *Vedic* and the *Upanishadic* perspectives, on the other hand, have a different philosophy and the survey of literature corroborates the fact that human treatment of and interface with the environment was dictated by the religious and the ethical set of values. This also constitutes a part of the Hindu cultural ethos, thereby underlining the Indian world view.

The *Vedic* literature (composed between 1600 and 600 BC) which is the oldest recorded literature in India celebrates the multifarious aspects of natural/ecological world in religious spirit. In *Vedic* literature, earth has been deified and endowed with all the motherly qualities. It is in the *Atharva Veda* that we find for the first time the elaborate description of lavishing appreciation on nature and environmental bodies. One does not lose sight of the unmistakable stamp of celebration in this scriptural text as the hymns invoke the mother earth:

> O Mother Earth! Sacred are thy hills, snowy mountains, and deep forests. Be kind to us and bestow upon us happiness. May you be fertile, arable, and nourisher of all. May you continue supporting people of all races and nations. May you protect us from your anger (natural disasters). And may no one exploit and subjugate your children. (*Atharva Veda*, XII.1.11)

What we observe here is that of the spiritual kinship existing between man and his surrounding nature. One of the significant pronouncements recorded in the *Vedic* text is the universal appeal/note, irrespective of nationality, gender, class. In Indian thought, the hierarchy between the human world the environment world is markedly absent, rather the attitude towards nature is one of reverence and worship. In sharp contrast to that, the present environmental hazards and the crisis aggravated by the utter indifference to and the merciless exploition of nature by human beings put our existence at stake. In this context the words as inscribed in *Vedic* texts are critique of modern human beings' insensible way of looking at nature from the colonizers' perspective by subjugating nature to our will. In them we hear the voice of the *Vedic* seers running across the whole of universe thereby emphasizing the vital and organic relation amongst all creatures. *Vedic* man's closeness to nature reminds us of the greek "Gordian Knot" which in the words of Vannucci:

> There is an interesting parallel of the Greek "Gordian Knot" in (RV. VIII; 40, 6), where Indra and Agni are invoked...may you cut the evils... like tangles of the creeping plant, the string to be cut is made of natural vines! The Vedic man shows his intimacy with nature as he sees health as the law and disease as the breakdown of order or law, which in this case is the break down of the equilibrium among internal factors of the body, of the mind. of their physiology as well as the breakdown of the balance between body and mind on one side and the environment on the other. (73)

Man always used to stay in a harmonious and non-reciprocal touch with nature. So much so that tree worship was greatly in vogue in ancient times. "Concerning flora in the Hindu religion as early as in the time of *Rg Veda*, tree worship was quite popular and universal. The tree symbolized the various attributes of God to the *Rig Vedic* seers ...Trees were considered as animate beings feeling happiness and sorrow. Green trees were likened to a living person"(qtd in Dwivedi 42), The *Rgveda*, which primarily dwells on the account of the immigration of the Aryans in the initial stage of the period, does not contain as much description of man-nature relationship as that of other *Vedas*. In the *Yajurveda*, we have beautiful description of nature:

> In spring the winds blow coolly like water, the rivers and ocean flow calmly and medicinal herbs are filled with sweet juice... In spring let trees give us sweet fruits, the sun physical strength and cows sweet milk. (*Yajurveda* XIII.27.9)

In the *Atharva Veda*, we have another view of the *Vedic* seers in constructing the relation of the environment not only to men but to all other creatures. The nature is not the sole territory of human beings; other biological creatures have equal right to the space of nature. The hymns articulate: "Born of Thee, on Thee move mortal creatures; Thou bearest them, both quadruped and biped. Thine, O Earth, are these five races of humans, for whom the Sun as he rises spreads with his rays, the light that is immortal" (*Atharva Veda*, XII, 1, 15). In another hymn, we have the idea that *Vedic* seers were committed to the welfare and benevolence of entire humanity including animal, and all the world of vegetation. This cosmic understanding provides us glimpses into their great concern for the "perception of the structure and dynamics of nature or what we would now call the ecological aspects of the many different environments colonized by man" (Vannucci 169). To quote another hymn from *Atharva Veda*:

> Now let the earth be peace and
> the air, the sky,
> The waters, the herbs, the trees be peace.
> Let the gods gle us peace and Peace
> comes from these calls for peace.
> With these peace calls that assuage everything &
> I give peace to the terrible here,
> To the cruel here, to the evil,
> Let them now be quiet, mean well,
> Let everything be good to us. (V,XIX, 9)

The Upanishads which constitute an important part of ancient Indian literature (mostly religious) also reflect on different aspects of nature. The creation of the Cosmos is also mentioned in connection with the elements of the earth. Here we have the knowledge that the whole cosmos is constituted of *Panch Mahabhutas* (Five Great Elements) which are mentioned in the *Aitareya Upanishad* as follows: "There exist the five elements: earth, air, space, water, and light" (*Aitareya Upanishad*, III, 3.). These fundamental elements are all that sustain and nourish the life of all human beings and after death they all assimilate into the universe again. These *Panch Mahabhutas* also go into making the Brahman. We have this reference in the *Shvetashvatara Upanishad*: "The Brahman by whom this entire universe is engulfed ... this creation is governed by Him as well as the five great elements: earth, air, space, water, and light" (*Shvetashvatara Upanishad*, VI, 2).

Like the *Vedas*, here also the vital and significant role of the trees is highlighted:

> Truly man is just like a tree. His hairs are the leaves and his skin resembles the natural bark. His blood streams

> forth out of his skin like the sap of a tree when he is cut... The flesh is comparable to wood, the sinews are like the inner bark, the bones are the inner core of the wood and the marrow resembles the pith of the tree.*(Brhad-aranyaka Upanishad* III.9.28.)

In, the *Chandogya*, we also have vibrant aspect of nature-worship. As in the *Vedas*, we have nature "in a meditative State". To quote Sharma, "... I would now like to share a phrase which makes some ecologists ecstatic. It says: dhyayativa prthivi:" "The earth, as it were, is in a state of meditation" (author's translation). "And the same verse asks us also to envision the atmosphere, the heavens, the waters, and the mountains as being similarly engaged." (55)

The question of pollution that has been plaguing the whole world for a long time was recorded in the ancient texts like Kautilya's *Arthasastra* and *Charaka Samhita*. At that time people had to abide by proper code of conduct in the society so as to maintain a healthy and balanced order in the society. On violation of these, they need to face punishment (qtd in Dwivedi 46). In Kautilya we observe:

> The punishment of one eighth of a pana should be awarded to those who throw dirt on the roads. For muddy water one-fourth pana, if both are thrown the punishment should be double. If latrine is thrown or caused near a temple, well, or pond, sacred place, or government building, then the punishment should increase gradually by one pana in each case. For urine the punishment should be only half. (qtd in Dwivedi 46)

The word "Vikriti" which is connotative of the modern system of pollution was very known to them. Charaka dwelt a great deal on the health hazards caused by pollution. To quote him:

> The polluted air is mixed with unhealthy elements. The air is uncharacteristic of the season, full of moisture, stormy, hard to breath, icy cool, hot and dry, harmful, roaring, coming at the same time from all directions, bad smelling, oily, full of dirt, sand, steam, creating diseases in the body and is considered polluted. (qtd in Dwivedi 46)

In the plays and poems of Kalidasa, the nature is radiated with spiritual aura. Its intrinsic relation to human life has time and again been emphasized in his poems. The images in his poetry are "determined by his conception of Siva's creative mystery" (Miller 7). To quote lines from Kalidasa:

> The water that was first created
> the sacrifice-bearing fire,
> the priest the time-setting sun and moon
> audible space that fills the universe,
> what men call nature,
> the source of all seeds,
>
> the air that living creatures breathe—
> through his eight embodied forms,
> may Lord Siva come to bless you.
>
> (quoted in Miller 7)

The unity of all creatures on earth recorded in *the Upanishad* is also voiced in Kalidasa. Nature is not merely a botanical construct, but it is much more than that. It is very much living and is symbolized as " the female half of his(Siva) cosmic unity" (Miller 8). In his plays, the space of the hermitage also underlines the idea of environmental ambience. The association of nature with hermitage as the macrocosm of the sylvan world has time and again been referred to in his plays. It points " a measure of serenity, purity and radiance of the All-being and is a countercheck against the impurity' hypocrisy and crookedness of court life" (Mishra 61).

In spite of great contribution of thought to the point of eco-human paradigm by Ancient Indian world view, it's important to remember that the present day discussion and writing of the environmental preoccupations in literature need to be treated in a different way from the one, the ancient Indian classical literature did. Today's sociocultural scenario is different and keeping that in view, the present day writers try to incorporate the question of politics, gender, in ecocritical context. But it is really astounding to note the ancient Indian people's concern with the environment and their sensible treatment of nature in the texts composed many centuries back. It would be futile to look for solution of the current environmental crisis in those texts, but what we can certainly look for is the moral and disciplined approach to the treatment of nature and the cultivation universal consciousness and values. It has to be kept in mind that the western view of segregating man from nature basically runs counter to the Indian, specifically Hindu ecological approach. Have we anything to learn from the Hindu view of ecology? It's very significant to note how in the contemporary environmental and ecological movements across India, the Hindu principles of ecology come in our rescue. Ramachandra Guha critiques vehemently the Non-governmental Organisations attempts at imposing the Western separatist principle in getting our ecological problems solved. He also opposes the Deep Ecology Movement's act of preservation and restoration of the polluted environment. Infact "Guha's critique is based on the Hindu perspective that people are not separate from but an interconnected part of nature. Therefore ecological goals are best served not by separating people from nature but by ensuring that people live in nature intimately in ways that support the diversity and harmony of the whole—which is God's body". Ancient Indian, specifically Hindu view of ecology and environment as recorded in the classical literature, was in the true sense of the term, much ahead of their time and it can be said without any modicum of doubt that the contemporary environmental literature or ecocritical literature has to get the impulse from the artistic and literary imagination/vision of the great classical masters of ancient Indian literature.

Works Cited

1. Chand, Devi. Trans. *Atharvaveda*. New Delhi: Munsiram Manoharlal Publishers, 1982. Print.
2. Coward, Harold. "Hindu Views of the Nature and Environment". *Nature Across Cultures: Views of Nature and the Environment in Non-Western Cultures*. Ed. Helaine Selin. Dordrecht, Netherlands: Kluwer Academic Publishers, 2003. 411-421. Print.
3. Deussen, Paul. *Sixty Upanishads of the Veda*, (2 vols). Translated from German by V. M. Bedekar and G. B. Palsule. Delhi: Motilal Banarsidass,1980. Print.

4. Dwivedi, O. P. "Classical India." *A Companion to Environmental Philosophy*. Ed. Jamieson Dale. Malden: Blackwell Publishing, 2003. 37-51. Print.
5. Miller, Babara Stoller. "Kalidasa's world and his plays". *Theater of Memory: The Plays of Kalidasa.* Ed. Babara Stoller Millller. New York: Columbia University Press, 1984.3-42. Print.
6. Misra, Vidya Niwas. "Man, Nature and Poet". *Creativity and Environment.* Ed. Vidya Niwas Misra. New Delhi: Sahitya Akademi,1992.56-62. Print.
7. Radhakrishnan, S. Ed. & Trans. *The Principal Upanishads*. New York: Harper and Brothers, 1953. Print.
8. Rama Pappu, S.S. "Humans, Animals, and Environment: Indian perspectives." *Philosophy, Humanity & Ecology: Vol. 1: Philosophy of Nature & Environmental Ethics*. Ed. H. Odera Oruka. Darby: Diane Publishing, 1996. 295-313. Print.
9. Sharma, Arvind. "Attitudes to Nature in the Early Upaniads". *Purifying the Earthly Body of God : Religion and Ecology in Hindu India*. Ed. Lance E. Nelson. New Delhi: D.K. Printworld, 2000. 51-60. Print.
10. Vannucci, M. *Ecological Readings in the Veda : Matter, Energy, Life*. New Delhi: D.K. Printworld, 1994.Print.
11. Worster, Donald. *The Wealth of Nature: Environmental History and the Ecological Imagination.* New York: Oxford UP, 1993. Print.

16

Life Lines: Water, Life, and the Indian Experience—Cultural Meanings, Social Significance and Literary Implications[1]

— Murali Sivaramakrishnan

My attempt here is to examine the socio-cultural significance and the imaginary implications of water—as *vital element, aesthetic metaphor, image* as well as *symbol*. So much like its fluidity and reflective transparency, this vital element, no doubt, plays a significant role in human imagination, and socially constructed reality alike. It conceals and reveals at the same time. Bereft of water there is no life, nevertheless, too much of it hastens human casualty. Drought and flood, famine, starvation and tsunami—all constitute the condition of human life on earth.

This essay is divided into two parts: the first is of a general kind that seeks to examine the social and spiritual significance of water in the multi-planed Indian context, while the second is of a textual kind focusing on a work of literature that aims to problematise cultural and historical meanings and the literary implications of water. The chosen text is a work of fiction.

Water has been endowed with vital significance all through human history. Civilisation and culture—however disparate they are—could be seen to be often so intimately linked to this dynamic fluid. From everyday ablutions to the highest symbolic levels the flow of water assumes infinite dimensions in every people's history. It cuts across human unconscious and the rational political and social systems alike.

Life in present day India is multifaceted and multi-layered, and often revolves round many planes at once- one could easily discern the profound relationship between religious beliefs and ingrained social structures and behavioural patterns. Even contemporary cinema and politics consciously or subconsciously veer in and out of the mythical and the *puranic*—often enough to the advantage of those in power and to the severe handicap of those underprivileged and over-exploited. Societal reality in present day India is reconstructed and deconstructed at will by the powerful memory of the past—however, often enough, only the hollow beliefs and superstitions survive as designs and counter designs of signification. The profundity of the Vedic poetic vision is now virtually unrecognisable in the macro world—however, it survives at the micro levels. Water, along with fire and earth is one of the elements that wield powerful religious and spiritual connotations-not merely in dominant Hindu ceremony and ritual but also in the other structures of beliefs and practice.

My intention is to close-examine the significance of this life-line-the trace, meaning and implication that water evokes as it flows in and out of the lives of people and places in India. Of course, I would also seek its non-human relevance and bearing.

▯

To the *RgVedic* poet, the elemental significance of water appears self explanatory. Sanskrit scholars have drawn attention to the fact that during the *Vedic* period which was affected severely by frequent floods, water naturally appeared to be the earliest element. Therefore in the poet's thought-process what was prior temporally, must have worked like the cause of the universe. No question is asked about the origin of waters. Even Vishwakarman—the creator of the universe, was preceded by them. In the *Brahmanas* water is described as co-eval with *Prajapati*. It was reasonable to *Vedic* people who saw land growing out of accumulations of river-torn silt that water was the primary element and source of all that existed.

The various theories of *Vedic* cosmogony could fall into three categories:

i) The most primitive ones begin with material principles

like water, earth, fire, ether and other elemental forms.

ii) Next come the abstract principles like chaos, time, night, desire, non-being (*asat*) and the like (however, asat only meant a sort of primordial non-differentiation).

iii) In the latest stage of development we come across divine principles like Prajapati, Brahman, Vishwakarma etc,—divisations of the energies of creation and maintenance, shaping and forming.

In all the *apah tatva* (water principle) plays a prominent role.[2]

The *Puranas* are codifications of myths and legends in the Indian traditions. The *Puranas* are referred to in the *Vedic* literature from the *Atharva Veda* downwards.[3] They embody mythical, moral as well as ethical values in their narrative content. In more ways than one the folk and the classical dimensions interpenetrate each other in the puranic texts. In one of the oldest *Purana*, the *Matsya Purana*, Parvati says:

> One who digs a well where there is little water lives in heaven for as many years as there are drops of water in it...[4]

Water also plays a significant role in the ceremonies and rituals—bathing in the river or any water body often signified moral, physical, and ritual purity.

Right from the ancient times in the Indian subcontinent, water thus has a profound physical and spiritual significance. Bathing in water-in a river, pond or back waters-means more than mere physical ablution-it also signifies moral and ritual purity. Sipping water from his cupped hands, three times, and allowing it to slip through tilted open palms, the Brahmin even today is expected to chant the hymns of the Vedic gods in ritual participation with the cosmic elements. Sprinkling water over one's head is also considered equivalent to the ritual bathing. As a rule, the Brahmin is supposed to take water in his cupped right hand and move it round the rice and victuals on the spread banana leaf before commencing each meal, chanting the various names of vital breath: *prana* (general term for breath in and out), *apana* (downward breath),*vyana* (the bond of union of the two), *samana* (common to both expiration and inspiration) and *udana* (leads the soul in deep sleep to the central reality). The presence of water thus is expected to cleanse and make everything sacred and pure. There is a widely held belief that before drinking even the dirtiest unclean water, if one were to chant the names of India's major rivers beginning with the austere Ganga—even any dirty ditch water will be freed from its contamination:

> *Gange cha Yamune chaiva Godavari Saraswati,*
> *Narmade Sindhu Kaveri jalesmin sannidhim kuru*
>
> (In this water, I invoke the presence of holy waters from the rivers Ganga, Yamuna, Godavari, Saraswati, Narmada, Sindhu and Kaveri.)

The Struggle for Water: Two Contemporary Instances:

Although in India water has thus mythical, religious, spiritual, metaphysical, symbolic, and physical significance/dimension, let us now look at two instances of the struggle for the physical aspect of water in contemporary India.

Narmada Bacchao Andolan: Save Narmada Movement

For over two decades now this struggle has come to symbolise the people's struggle for a just and equitable truly democratic society in India. Narmada is among the great rivers of India, spreading over a vast region in the west, for the most in the state of Gujarat. The government planned to build 30 large and 135 medium and 3000 small dams to harness the swift flowing waters of this mighty river and its tributaries. The projected purpose was of course to provide larger amounts of water and produce large amounts of electricity for the purposes of development.

Opponents of the dam question the basic assumptions of the Narmada Valley Development Plan and believe that its planning is unjust and iniquitous and the cost-benefit analysis is grossly inflated in favour of building the dams. It is well established that the plans rest on untrue and unfounded assumptions of hydrology and seismicity of the area and the construction is causing large scale abuse of human rights and displacement of many poor and underprivileged communities. They also believe that water and energy can be provided to the people of the Narmada Valley, Gujarat and other regions through alternative technologies and planning processes which can be socially just and economically and environmentally sustainable.

Coca-Cola in Kerala

Now for another instance of physical contamination of the sacred elixir: Kerala is the southernmost state in India and it is famous for its coconut palms, greenery, lagoons, water ways-three major backwaters (lakes)-mountains (the Western Ghats), and rivers (forty odd, west-running). It is also famous for democratically electing the first ever communist ministry in the world in 1956. Over the last decade Coca-Cola had managed to procure large areas of land with adequate fresh water sources and set up their plant in a place named Plachimada in Perumatty Panchayat in the Palakkad District.

Meanwhile out in the real world of the Indian countryside, Coca-Cola's bottling plants were getting less enthusiastic reviews. Coca-Cola had sound reasons in zoning in on Plachimada. A rain-shadow region in the heart of Kerala's water belt, it has large underground water deposits. The site Coca-Cola picked was set between two large reservoirs and ten meters south of an irrigation canal. The ground water reserves had apparently showed up on satellite surveys done by the company's prospectors. The Coke site is surrounded by colonies where several hundred poor people live in crowded conditions, with an average holding of four-tenths of an acre. Virtually the sole source of employment is wage labour, usually for no more than 100 to 120 days in the year.

> Ushered in by Kerala's present "reform"-minded government, the plant duly got a license from the local council, known as the Perumatty Grama panchayat. Under India's constitution the panchayats have total discretion in such matters. Coca-Cola bought a property of some 40 acres held by a couple of large landowners, built a plant, sank six bore wells, and commenced operations.

> Within six months the villagers saw the level of their water drop sharply, even run dry. The water they did draw was awful. It gave some people diarrhea and bouts of dizziness. To wash in it was to get skin rashes, a burning feel on the skin. It left their hair greasy and sticky. The women found that rice and dal did not get cooked but became hard. A thousand families have been directly affected, and well water affected up to a three or four kilo meters from the plant. The old village wells had formerly gone down to 150 to 200 feet. The company's bore wells go down to 750 to 1000 feet. As the water table dropped, all manner of toxic matter began to rise too, leaching up to higher levels as the soil dried out. Further the waste from the bottling plant was taking the form of the toxic sludge, along with profuse daily donations of foul wastewater.
>
> The company told the locals the sludge was good for the land and dumped loads of it in the surrounding fields and on the banks of the irrigation canal, heralding it as free fertilizer. Aside from stinking so badly it made old folk and children sick, people coming in contact with it got rashes and kindred infections and the crops which it was supposed to nourish died.
>
> Lab analysis by the Kerala State Pollution Control Board has shown dangerous levels of cadmium in the sludge.
>
> But now the Coke officials are under pressure. So is the Kerala government. The government has now separately sent samples of the sludge to various laboratories to test whether the fertiliser indeed contains poisonous toxic materials.[5]

Contradictions: As we can see the lived experience in India is so full of such contradictions. The greatest and the most revered river in India is the Ganges. It is the mythical purifier-but the most polluted river. This is the case with most regional rivers and lakes-they may be so very highly venerated as sources of life and spirit but polluted drastically. Our Indian reality is so much multilayered.

Ashokamitran's Water[6]

When the first part of my essay sought to understand the multiple planes and manifold significance that the elemental sign of water would evoke, the second part would be a re-reading of a work of fiction—Water—originally written in Tamil by Ashokamitran.[7]

This work is a sort of social documentary that unfolds over a period of time in a small village in South India. As the name would very well suggest, the entire theme of the work revolves round the image of water—real as well as spiritual and imagined. Even though the theme chosen could easily overflow its fictional boundaries, and end up even as a mere "chutnification of history,"(the phrase of course is Salman Rushdie's) the writer has dexterously contained his material in an artistic manner focusing for the most on a specific locality and streamlining his tale to a fine narrative point. The story flows in fits and starts. After all, a tiny streamlet could also exhibit the force and ferocity of a tidal wave.

In one of our many personal conversations the writer told me that he liked to talk about the life he has seen and experienced-urban and middle class. For him technique is what surfaces automatically—there is no straining after form and style. The story does it for him. Incidents might appear to be loosely strung together in his narratives, apparently unconnected and unrelated, however, they form a deeper chain of significance and meaning because the life of middle class India is so rich in contradiction and incoherence.

Water, is no different. It is an intense and poignant portrait of a suburban corner of Chennai city with its un-motorable torn up streets, dusty and filthy, with its teeming population. There are good people and nameless bad people, uncaring governmental activities, depraved individuals like Bhaskara Rao who mercilessly exploit innocent and willing victims like the heroine Jamuna. There are also more determined individuals like her sister Chaya—married to an army person who cannot manage to get a transfer to the south, and who also has a son she has to leave in the care of her other unwilling relatives.

The main narrative is centred round the lives of these two sisters, and their struggles to live in the drought ridden city, while all around them the saga of an entire humanity is portrayed.

The background of the novel is the severe drought that hit this area in 1969. However, there is no direct reference to this historical fact in the book. The story revolves round the lives of the two sisters-Jamuna and Chaya—and their lives-urban and lower middleclass. The backdrop is the Chennai city, and the theme for the most is water. Like all rivers in India which originate from the spiritual heights of misty mountains or hill ranges, only to flow down through the mundane middleclass city-scapes or lifeless villages or polluted farmlands, and finally empty into either the Arabian Sea or the Bay of Bengal—polluted and dirtied, carrying all the debris and impurities of this world, and yet even though disfigured are supposed to retain some significant element of the original source-the characters in the work pass through life's hard, grim, realities in a semi-dreamlike state and emerge into the elemental real at the end of the narrative, more focused and more determined than ever. When Jamuna and Chaya walk away hand in hand leaving Bhaskara Rao's car stranded in the mud of the streets, they are in fact proclaiming their new found meaning, strength and determination. Water, of course, always finds its own level.

Water is the major theme of the novel-or rather the lack of it at the beginning, and after the arrival of the rains, the mud and squelch of the sub-urban middleclass living spaces.

Ashokamitran has said that he wrote *Tannir* (the original Tamil title of *Water*) as a long short story beginning only with the picture in his mind of a girl who was always seen to be carrying a huge water pot, whether it was at two in the morning or in the heat of the day at high noon. It was serialised in *Kanayaazhi* from July to November 1972 and published in book form in 1973. Many critics in Tamil have drawn attention to the multifaceted levels of meaning in the work, including the symbolic overtones in the title. In her note the translator points out that the story describes the daily struggle of ordinary men and women to find enough water for their needs, their ingenious strategies, their ability to search out any source, and their interaction with an inept municipal staff which is incapable of providing the infrastructure necessary for a growing urban population. The discrete images of life in the sweltering city streets are strung together as in a film. They are unified in the narrative as the winding street itself and the image of water.

Jamuna and Chaya are the main characters. So is Teacher Amma who symbolises the will to live under all odds-veering between a nagging household and a gritty career-and she steps into Jamuna's life as a hot breeze. Bhaskar Rao stands for the developing city's debilitating force that draws everything into its sordid vortex and all but sucks Jamuna in it too. The opening lines of the novel signal the significance of events to follow:

> Jamuna had been lying in bed for the past half hour listening to the water pump. As soon as she detected the slightest change in its sound, she instantly shook herself awake and, switching on the light, she picked up the two brass water pots and ran downstairs.Water pumps were working away in all the houses of the neighbourhood. At that early morning hour, you could hear the noises unique to each and every water pump, both separately and in unison. They sounded in all the streets, in this one and the next one and the one next to that, in each of the houses there.

The entire narrative revolves round the imagery of water. The simple pressures of daily living and the higher spiritual levels of symbolic ablutions are washed over by this fluid image. The city's bustle, heat and dust, weave in and out of Jamuna's consciousness. While her more liberal minded sister Chaya moves about in a more unfettered manner, Jamuna fumbles along not too keenly aware of where and in which direction life is leading her. Bhaskar Rao, the small time cinema producer, a self-seeking lout to the core, exploits Jamuna in her more abstract moments, whisking her off as an exhibit to please his clients.

Albeit a trifle reluctant Jamuna gives in to all his whims and fancies and feels abysmally ashamed after every escapade into the city's entrails with him. When Chaya leaves her in a fit of anger she feels her absence as a deep void that makes her life even more meaningless. Even the fetching of water she used to enact as a daily routine and ritual becomes devoid of any significance. Teacher Amma then takes the lead and eggs her on to fetch water in her company.

> "...come on, bring your water pot. I will get you some tap water as well."
>
> "I have enough, Akka."
>
> "Do you mean the water from the street tank?"
>
> "Yes."
>
> "If you keep it for another day it will be really stale."
>
> Jamuna switched on the lights.
>
> "Hurry up," said Teacher Amma.
>
> "I really don't need any water today, Akka. My water pot is still half-full."
>
> "I,m telling you to come along. Is that amount of water going to remain the same always? And what will you do if a lizard falls into it suddenly?" (pp.46-47)

This contact exposes Jamuna to Teacher Amma's family life. She although a sensitive woman, had been married at a very young age to a congenital consumptive and is constantly ill treated by her sick mother-in-law. Despite her miserable and wretched condition she is able to instil a certain amount of determination into Jamuna's failing will. Jamuna survives a futile attempt at hanging herself and continues her quest for water. In the meanwhile the life of the streets changes—workers dig up the sides of the road in the pretext of installing a water tank and later rake up the entire street in order to reinstall new water pipes. The incoherence and disorder of unplanned metropolitan chaos is reflected well with in the inner life of the protagonist. Or perhaps, it could be read both ways—the chaos is inside first. There are sparks of wit and humour that trigger certain minor incidents which bring considerable relief to the other wise direct narrative. In all the while the focus does not shift from water or the lack of it.

There is a distinct space for the spiritual quality of water, that still retains its imaginary hold in the mind of some people-at least the, so-called upper class Brahmins. Consider this incident that Ashokamitran chooses to highlight:

> The people collecting the water were going up and down the street as if they were part of a procession.at any given moment, all these people were somehow able to get enough water to run their households. It looked as if it was during this time of drought alone that these people used water most plentifully. Yet in actual fact, in normal times they probably used much more water than they did now, simply because water had never been a commodity to be used with such care. Jamuna stood there, watching the street, not thinking about herself anymore. A Telugu Brahmin widow wearing a red sari, quite an old woman, turned into the street, carrying a large water pot on her hip. She was probably carrying that much water for her use alone. Heaven knew how far she had been carrying that pot. But after she came within Jamuna's vision, it turned out that it was not going to be possible for her to totter even ten steps further. She stood still and looked as if she was swaying. Then, dropping her water pot, she fell right to the ground. Jamuna ran up to her and attempted to pick up the water pot. The old lady cried out, "Don't touch it. Don't touch it." Jamuna held on to the old lady and helped her sit up. Still somewhat unsteady despite Jamuna's grip on her, the old woman kept falling over. Even so, she reached out a hand towards the water pot which now lay on its side, slightly dented, in a pool of muddy water. Jamuna held on to the old woman more firmly. By that time, four or five others had gathered there. The old lady pulled herself together after a fashion. But before she could herself straighten up the water pot which lay next to her, someone had pushed ahead, touched it and lifted it up. The old lady was immediately overcome with weakness once again. When those who had gathered there spoke to her kindly and tried to help her to her feet, she even began to cry. She rose unsteadily to her feet and grasped hold of her water pot, barely able to stand. She tipped out its contents-for it was nearly half-full-on to the ground. She shook herself free from Jamuna's grasp and tottered forward on her own. (pp.80-81)

Jamuna appears bewildered at the Brahmin widow's behaviour. Even through the agonising days of drought many believers in orthodoxy, practise what is known as madi or ceremonial purity. For the Brahmin, the early morning bath is no less significant in the spiritual as well as the physical planes. Cleansing the body is also at the same time a process of cleansing the mind and spirit—the *antaratman*. The water for this purpose has to be collected by the ritually pure person and should not be touched or breathed upon by the impure other. In case it is thus contaminated by the touch that water becomes unusable. The old woman would rather pour out the rest of the impure water than carry it to her home. Perhaps Ashokamitran works out this irony in the narrative by setting this scene as a deliberate backdrop against which the actions and decisions of Jamuna, Chaya and Teacher Amma could be examined. All the people who inhabit the novel are victims of the development syndrome. Their lives, moral as well as ethical choices are challenged and threatened by the absence of free flowing water and their ever narrowing suburban spaces. As they are crowded out, their beliefs also are tested. Even when it finally drizzles they do not have much option to collect and retain the rain water and are forced to watch it flow down to the streets and collect in puddles and squelchy mud.

For Jamuna the regular collecting of water seems to provide some meaning to her otherwise meaningless life—it is her major occupation. She is aware of water—or rather the lack of it always. For Chaya, her sister's obsession with water seems only to signify her lack of commitment to any other action. She is a more determined person, although at many places along the narrative she too breaks down and seeks the comfort of her sister's shoulders to cry on.

At the end of the novel, Bhaskar Rao's new car is stranded in the muddy streets, and he stands around quite helplessly, Chaya takes Jamuna's hand and the sisters move on unconcernedly, yet certainly more determined to face the odds of life. Throughout the story the tension is built up and maintained by the patched up life in the streets that thirst for water, and in the end the mud that remains after the rain washes over.

Water for Jamuna symbolises all her hopes, desires and aspirations. It also signals her metaphysical existence although she constantly seeks it is its physical form. Very much like the parched and thirsty land her soul searches for the elixir of life in all eagerness and hunger. Her dreary existence on the other hand corresponds to the dreariness that she sees all round her. The barrenness of the suburban existence stands for her own meaningless existence—physically and philosophically. The only work that she does is the daily collecting of water in containers from taps and distant wells. And much like the substance that water is, she allows herself to be contained and manipulated by the others. All this she does with undisguised disgust and uncommitted will. Ultimately when it does drizzle after a long spell of drought she finds that she has nothing more to look forward to. Bhaskara Rao has left and so has Chaya. Her own family had nothing more to offer her. Nevertheless the downpour symbolises a new direction in Jamuna's life and in the life of the streets as well. Bahaskara Rao's new car gets stranded in the muddy streets and metaphorically allows Jamuna to get out of his reach finally.

Water also makes and remakes the thread of continuity for the governmental activities in the novel. The authorities begin by digging the roadsides during the drought in search of the water pipes that were laid earlier. The unskilled workers manage to leave the road in a ramshackle state afterwards, stranding vehicles and the public after a few showers. Further they also inadvertently break up the electrical connections of the streetlights plunging the streets into darkness for months. The helpless life of the middle class that are even more complicated by the meshes of the establishment and its power structures are clearly highlighted by the narrative. And water forms its connective thread.

Ashokamitran's *Water* is a tale that does not consciously strive to please the reader. It depicts certain characters in certain situations in an impressionistic montage of happenings much like the perennial imagery of water that holds it together. All characters are linked up either by the scarcity of water or its lack and dire need—at many levels—the gross physical, the metaphysical and the spiritual.[8]

In the rhetoric of Indian views of reality as lived experience the significance of water is thus multifaceted. Water is both transparent and opaque—it conceals and reveals at the same time. Bereft of water there is no life, however, too much of it hastens human casualty. Life depends on fresh water and the regular water cycle. If the water sources are over exploited severe drought is the result. On account of constant misuse and unwarranted developmental activities desertification has set in where once the vital fluid flowed liberally replenishing and nourishing the dynamic of existence. Despite all the scientific knowledge that the growth and extensive advancement and reach of Ecology as a scientific discipline has effected and ensured,—environmental education is compulsorily introduced even at the lower schools—technological wastage continues unabated in India. Primarily, because the powers-that-be, act in deliberate unconcern, indifference and incoherence. Lack of proper planning and uncaring misuse of available infrastructure and resources, coupled with managerial clumsiness, account for the most. As a developing country India has in the present become a vast unfolding field of possibilities and potentialities—a veritable market place. Economic forces that manipulate the gator like grip of market capitalism appear to have unleashed their terrorist activities in full swing. Everywhere we come across the slogans of development and urbanisation. Of course, that is apparently the only way open before any country in the present. Indigenous ways of development and calls to appropriate technology however, simply fall on deaf ears.

The twenty first century is branded and sold out as a century of information and hitherto unprecedented rise in the mass production of consumables and luxury items: the ill-informed common man and woman are swept off their feet by the onrush of this materialistic culture. Capitalist forces present the promise of collective satisfaction and plenitude of goods and goodies albeit through a virtual image. Like the donkey in the fable that keeps on moving forward in the hope of reaching the carrot dangling within its easy reach in front of it, we move on without realising that the carrot is attached to a stick tied to our necks!

Water is one of the prime elements that our lives depend upon, and fresh water is over-exploited and misused. Our uncaring ways of life selfishly motivated and technologically never gratified, wanting more and more but having less and less, lead us on to our inevitable doom. The environment is not unlimited; nature is not unlimited for the taking. We pollute our earth, pollute our air and pollute our waters. We create debris and garbage on an unprecedented scale and are left gaping on how to dispose of all the waste. Industrial waste as well as nuclear waste, would add to the growing concern of any responsible, thinking intellectual in. Sadly enough, Bhopal and Chernobyl seem to have left hardly any remnants of global concern in our minds. After all human memory is so short!

The fact remains that, if at all the world is to go in for another war, it is certainly going to be for the dominant rights and privileges over safe and uncontaminated drinking water. While on one side we in India look upon water as a sacred element—whatever our religious bias and beliefs—we also relentlessly misuse this elixir of life inadvertently—the ignorant masses perhaps unknowingly towing the line that the developed countries throw in. When we suffer excruciating droughts and periodic failures of the monsoon cycle, we also reel severely under unforeseen and untimely floods, cyclones and tsunamis. At every step the natural cycle of water speaks dread, death and needless loss to the common people of this part of the world. While the rich and the politically powerful somehow mange to escape or struggle free from the natural and unnatural disasters, the poor and the marginalised suffer terribly. The division between the haves and the have-nots speak volumes here as elsewhere. Water thus has a social-political dimension that needs to be re-examined, because the possession, reach and accessibility to clean potable water constitute its political dimension. After all, water is one of the prime elements in the making and unmaking of life and Water is our life line.[9]

The ancient Indian experience of undivided existence that does not bifurcate the material and the spiritual is beautifully brought out in the *Purna mantra*:

Purnamadah Purnamidam Purnatpurna mudachyate.
Purnasya Purnamadaya Purnamev avashisyate.

(That is Purna, this is Purna, Purna comes out of Purna.
If Purna is subtracted from Purna, still Purna is left.)

—(*Yajurveda, Shanti Mantra*—Chapter 40)

This mantra could be read as a celebration of the benignity of the earth and its biological and metaphysical diversity and plenitude. Each of the *pancabhutas*—the five elements—individually are complete in themselves; they symbolise a holistic existence, a completion—sampurna—within and without. Water is one among the primal and complete elements. We can use it, we can misuse it, we can celebrate it. Somewhere along the way when we ceased to look upon the non-human nature as sacred, we allowed ourselves to look upon it as mere resource to be utilised and exploited. And that has made all the difference.

To conclude, what better than this *Universal prayer?*

Sarveṡām Svastir Bhavatu
Sarveṡām Sāntir Bhavatu
Sarveṡām Pūrnam Bhavatu
Sarveṡām Mangalam Bhavatu

(May good befall all,
May there be peace for all,
May all be fit for perfection, and
May all experience that which is auspicious.)

Sarve Bhavantu Sukhinaha
Sarve Santu Nirāmayaha
Sarve Badrāni Pasyantu
mā Kascidh-dhuhkha Bhāga-Bhavet

(Om, May all be happy.
May all be healthy.
May we all experience what is good and let no one suffer.
Om, Peace, Peace, Peace!)

Notes

The state of Tamil Nadu lies on the east of the Western ghats—the row of mountains that stretch from Mumbai to Cape Comorin in the west coast— and is dependent on North East monsoons for rain water. (There are two cycles of Monsoons—the South West that usually begins in June and the North East that sets in by October.)It is usually draught-prone and yet in the last monsoon cycle there were unprecedented floods. There are countless reasons for the droughts and floods but the most dominant ones are merciless and relentless deforestation, unwarranted, unplanned urban developmental activities on river banks (and beside other fresh water sources) that cause siltation, sedimentation and soil erosion, etc.

Purna is derived from root "*Pri*" meaning to fill. Thus *Purna* means full or complete. In this context, *Purna* can also be interpreted to mean infinity.

References

1. A pre-revised form of this essay was presented at the Second Biennial Conference of EASLCE in Austria— "Water: Literary, Cultural and Environmental Perspectives" from 28th April-1st May 2006. As far as possible I have also tried to incorporate the various aspects of the debates that ensued in a reworked form in this essay.
2. In the early Vedic cosmogony, the *apa-tattva* receives a good deal of attention. This was prompted, probably, by the historical background of the period. Historians have pointed out that frequent floods in the Indus Valley have influenced the thought-process of Vedic Aryans when they stated that water was the primordial substance out of which the Universe came into being (*ÎV,* 10.129.1). The Indus Valley appears to be a target of frequent floods. So also its extention viz. Lothal, Rangpur-Koth in Saurashtra. The accumulation of flood-debris at various sites is an evidence for the inundation of Mohen-jo-daro. The 12 meter-high hill of silt at Budh Takkar is the result of a great flood of long duration, almost a deluge, which must have turned the lower Indus Valley into a vast lake. The Carbon-14 dates for the two great floods at Lothal are 2015 + 115 BC and 1900 + 115 BC. The flood debris of the latest level of HR mound at Mohen-jo-daro is assigned to 2000 BC. The flood in *circa* 1900 BC assumed such fierce proportion and was so prolonged and devastating that all the Harappan settlements in Saurashtra, Kutch and S. Gujarat were wiped out.

3. See J N Farquhar, *An Outline of Religious Literature of India*, 1920; rpt. Delhi: Motilal Banarsidass, 1984.
4. Vasudha Narayanan, *Water, Wood and Wisdom: Ecological Perspectives from the Hindu Traditions*, Daedalus, Fall 2001,Vol.130, No 4,pp 179-206.

5. Courtesy: http://www.counterpunch.org/cockburn04162005.html
6. Ashokamitran, Water, translated into the English by Lakshmi Holmstrom, (New Delhi:Katha,2001)
7. Ashokamitran is the pseudonym of Tyagarajan, who writes in Tamil and lives in Chennai, the capital city of Tamilnadu, south India. Ashokamitran belongs to the eminent group of Tamil writers who won recognition for their storytelling ability and craftsmanship in the 1960-s. He began his career with the prize winning play *Anbin Parisu* which were followed by many short stories and a volume of stories—*Viduthalai*. He has authored eight novels, including *Karinda Nizhalkal, Padinettavadu Atchakodu, Indru, Manasasrovar, Vizhaa Maalai Podil*. He is also a distinguished essayist and critic, and was the executive editor of Kanaiyaazhi for many years. Ashokamitran received the Sahitya Akademi award in 1996.

Works Cited

Ashokamitran, *Water, translated into the English by Lakshmi Holmstrom*. New Delhi:Katha,2001.

Bhise,Usha R. Water Element in Vedic Cosmogony, Prakrti, Vol 2.*Vedic, Buddhist and Jain Traditions*, Edited by Sampat Narayanan. New Delhi: IGNCA.1995.

Farquhar, J N. *An Outline of Religious Literature of India*, 1920;rpt. Delhi:Motilal Banarsidass, 1984

Maslin, Mark. *Global Warming: A Very Short Introduction*. Oxford: Univ.Press,2004.

Narayanan, Vasudha. *Water, Wood and Wisdom: Ecological Perspectives from the Hindu Traditions*, Daedalus, Fall 2001,Vol.130, No 4, pp 179-206)

Vandana Shiva et al. *The Future of Progress: Reflections on Environment and Development*, Dehra Dun: Natraj Publishers(Indian Edition), 1994.

Radhakrishnan. S. (ed.)., *The Principal Upanishads*, London: George, Allen &Unwin, 1953.

Contributors

Murali Sivaramakrishnan poet, artist and literary critic—is Professor and Head of the Department of English, Pondicherry University, Pondicherry, India. A specialist in Indian aesthetics and literary theory, he is a scholar of international repute deeply concerned with indigenous values and environmental issues. He has authored numerous articles on art, literature, and ideas in acclaimed journals and has also coordinated seminars and conferences in the field of literature, eco-aesthetics and literary theory, postmodernism and cultural studies. Murali has travelled widely and participated in significant international conferences in Jamaica, Spain, Taiwan, Austria and Brussels, England, and lectured in the University of Salamanca, Valladolid, and Extremadura in Spain, the University of Umea in Sweden, and Pembroke College, Cambridge, UK. He has held two solo exhibitions of his works in Germany and many in India. Dr. Murali has as much involvement with nature and environment as with painting and poetry. He was also Fellow at the Nehru Memorial Museum and Library, Teen Murti, New Delhi, and Associate of the Indian Institute of Advanced Study, Shimla. He is member and coordinator of research of the *Herman Hesse Society of India.* Dr Murali Sivaramakrishnan is the founder President of ASLE India (*Association for the Study of Literature and Environment) and* was a *Fulbright* awardee in the University of Nevada at Reno (2006-07). He was invited to read his poems as part of the inauguration of the International Conference on Poetic Ecologies, held in the *Universite Libre de*

Bruxelles, Belgium, in May 2008. Murali's sculpture work (cast in fibre) of Prof. C.D. Narasimhaiah now adorns the conference hall of *Dhvanyaloka,* Mysore. S Murali's academic publications include: *Tradition and Terrain: Aesthetic Continuities*. New Delhi: B R Publishers, 2005 (co-authored with Usha V.T.); *Nature and Human Nature: Literature, Ecology, Meaning* (New Delhi: Prestige Books, 2008); Culture *and Image: The Dynamics of Literary, Aesthetic and Cultural Representation* (New Delhi: Atlantic, 2010); Learning *to Think Like Myself* (New Delhi: Gnosis, 2010); Cries *and Whispers: Nature, value, Education* (Essays—forthcoming); *Sri Aurobindo's Aesthetics and Poetics: New Directions* (Essays—forthcoming).

Mihai A. Stroe is Associate Professor at Bucharest University, Faculty of Foreign Languages and Literatures. In 2003 he received a PhD in Philology from Bucharest University. He is a specialist in English, American and German literature; comparative romanticism and archetypology; the theory of paradigms in literature, science and religion; and general poetics. The recipient of several scholarships and grants (Vienna University, 2000-2001; Danish Cultural Institute, 1994), Mihai A. Stroe is also a laureate of the Theodor-Körner Prize for the Promotion of Science and Art (Vienna, 2001). He is, moreover, a member of The New York Academy of Sciences (NYAS) and a Postdoctoral Fellow of Yale University. He has published over 50 articles, authored several books and translated a number of Romantic poets into Romanian. His works include:

— Friedrich Hölderlin, Hyperion—Un fragment (1998) (bilingual critical edition);

— William Blake, Cei patru Zoa (1998) (bilingual critical edition);

— Romantismul german si englez (2004) (the augmented version of his PhD thesis: German and English Romanticism—The Science of Archetypes, the Hypothesis of Interfinitude and the Golden Section);

— the series William Blake, Opere Complete (Complete Works; bilingual critical edition, projected in ten volumes): Vol. 1: Vala (2006); Vol. 2: Milton (2006).

— John Clare, Poeme (2007) (bilingual critical edition);

— Geneza: Dicþionar englez-român (Genesis: English-Romanian Dictionary) (2008, forthcoming). Currently, Mihai A. Stroe is working on an interdisciplinary Fulbright project entitled *Paradigm Shifts in Literature and Science* under the scholarly advisorship of Prof. Harold Bloom (Yale University) and supervision of Prof. Paul Fry (Yale University).

Reena J Andrews is currently Selection Grade Lecturer in the Department of English, St. Paul's College, Kalamassery, affiliated to Mahatma Gandhi University Kottayam, Kerala. She took her post graduate degree in English from St. Joseph's College, Devagiri, Calicut with the First Rank in the University. She was also the first rank holder in M. Phil from the Department of English, University of Calicut. She has a Post Graduate Certificate in the Teaching of English from the *Central Institute of English and Foreign Languages*, Hyderabad. She was awarded Ph.D. by Mahatma Gandhi University, Kottayam in 2004 for her work on the poetry of Robinson Jeffers. She has completed a research project on the poetry of Keki N. Daruwalla.

Bajrang S. Korde is currently Professor in the Department of English, University of Pune. He has travelled widely and lectured on literature and literary studies. His field of interest is British literature.

Margarita Carretero-Gonzalez is Senior Lecturer in English Literature at the English and German Department of the University of Granada (Spain). Her interests include women's studies, cultural studies, children's and fantasy literature, and ecocriticism as a tool for literary analysis. She has published extensively, delivered lectures and participated in round table discussions on ecocriticism and ecofeminism. Her most recent publications in the field of ecocriticism include articles on J.R.R. Tolkien ("Mujeres de sangre, mujeres de savia: aspectos sociales y biológicos en la construcción de los géneros en The Lord of the Rings", in De habitaciones propias y otros espacios conquistados: estudios sobre mujeres y literatura en lengua inglesa en homenaje a Blanca López Román. Eds. Margarita Carretero-González, María Elena Rodríguez-Martín & Gerardo Rodríguez-Salas. Granada: Editorial Universidad de Granada, 2006; "When Evil Responds to Evil Practices: a Warning from the Ents of J.R.R. Tolkien's Middle-earth", in Truth, Evil and Reconciliation. Ed. Margaret Sönser Breen. Amsterdam & New York: Rodopi, 2004: 149-165), on C.S. Lewis ("Sons of Adam, Daughters of Eve, and Children of Aslan: An Environmentalist Approach to The Chronicles of Narnia", in C.S. Lewis: Life, Works, and Legacy (Vol. 2: Fantasist, Mythmaker and Poet). Ed. Bruce L. Edwards. Westport, (Conn.) y Londres: Praeger Publishers, 2007: 93-113) and Virginia Woolf ("'I Am Nature's Bride: Orlando and the Marriage of Nature and Culture". In-Between: Essays and Studies in Literary Criticism. Vol. 14, no. 2, 2005: 259-268). She wrote her Ph.D. dissertation on J.R.R. Tolkien's the Lord of the Rings (Fantasía, épica y utopía en The Lord of the Rings: análisis temático y de la recepción. Granada: Servicio de Publicaciones de la Universidad de Granada, 1997) and she is currently working on a biographical study on Mabel, Edith and Priscilla Tolkien, as well as on a book exploring the relationship between women and the environment in the works of British writers from Wollstonecraft to Woolf.

Priyadarshi Patnaik (b. 1969) is a creative writer, painter, translator and photographer. A number of his poems and short-fiction have appeared in various journals outside and in India including Ariel, Oyster Boy Review, Hudson View, Melic Review, Still, Toronto Review, Kavya Bharati, Indian Literature and Muse India. His translations and critical writings on translation have appeared in Translation Today, Visva-Bharati Quarterly, Muse India and many edited volumes. He has published two anthologies of poems, a critical work on Indian aesthetics and co-edited two volumes on Aging and

Dying (Sage) and Time in the Indian Context (D K Printworld-in Press). He is presently editing a volume on Orissan Medieval Poets and writing a monograph on poet Achyutananda for Orissa Sahitya Akademi. Patnaik is currently Associate Professor at the Department of Humanities & Social Sciences, IIT Kharagpur, where he teaches literature, communication and visual aesthetics. His research interests include Indian aesthetics, media & multimedia studies, visual & nonverbal communication, and translation.

Rohith. P. is currently Lecturer in the Department of English, Deen Dayal Upadhyaya College, Delhi University. His doctoral dissertation submitted to the Centre for Comparative Literature, University of Hyderabad, was on literary/artistic reactions to the campaign to save the tropical rainforest in the Silent Valley in Kerala. Rohith's areas of interest apart from Ecocriticism are cultural studies, critical disability studies and comparative literature. He has contributed articles on ecological criticism to various anthologies and has presented many research papers in this field.

Elmar Schenkel is Chair of British Literature at the University of Leipzig /Germany. He has published essays on bicycles, volcanoes, alchemy and architecture, as well as on Nietzsche, H.G. Wells, Chesterton, Joseph Conrad and John Cowper Powys. He is also a writer of travel books on India, Japan, the US and Russia and has published poetry, essays, short stories and two novels. Forthcoming: a book of aphorisms and of children's stories.

Usha V.T. is currently Associate Professor and Head at the Centre for Women's Studies, Pondicherry Central University, Pondicherry, India. Her publications include books and scholarly articles in the field of feminist theory, women's discourse, women and media. She did her post graduation in English Literature and specialized in British poetry for her PhD. Her postdoctoral research moved toward the emerging areas of women-centred discourses. Under her eminent guidance and direction the Centre for Women's Studies in

Pondicherry University has produced a large number of dissertations and research papers. She has travelled widely in India and abroad and lectured and made presentations in several academic forums as well as nonacademic bodies. She has organized a large number of seminars and workshops particularly relating to gender and Women's Studies. She has been an Associate of the Indian Institute of Advanced Study, Shimla and is currently an active member of the Curriculum Development Committee for Women's Studies of the University Grants Commission. Dr Usha is member of the Board of Studies for Women's Studies, and has also served in the capacity of member of advisory committees in many universities. Dr Usha's publications include *Modern English Literature— The Real and the Imagined: The Poetic World of Ted Hughes* (Jaipur: Mangal Deep, 1998); *Figuring the Female: Women's Discourse Art and Literature.* (New Delhi : The Woman Press, 2006) *Tradition and Terrain: Aesthetic Continuities* (New Delhi: B.R. Publishers, 2006) [both co-authored with Dr. Murali Sivaramakrishnan]; *Women and Moral Policing* (Pondicherry: Centre for Women's Studies, Pondicherry University 2009); *Gender based Violence: Some Primary Facts* with Amrutraj R.M. (Pondicherry University Centre for Women's Studies, 2009).

K. Reshmi, is currently Assistant Professor, Department of English, Pondicherry University, Pondicherry. Her doctoral degree was in the field of Ecocriticism. She has published articles on ecocritical and feminist writings in edited anthologies, books and journals. Reshmy's interest also includes the fields of Commonwealth Fiction and Comparative Literature.

Sagar Mal Gupta is presently Chair of the Department of English and Humanities at Sobhasaria Engineering College, Sikar Rajasthan. Prof Gupta earned his MA and Ph.D. in Linguistics from the University of Hawaii, USA. He did his PG Diploma in Applied Linguistics from Edinburgh University, UK. He completed his Diploma and MA in Distance Education from IGNOU, New Delhi. S.M. Gupta has 45 years experience of teaching English Language and Literature in India and abroad (over two years at the University of Aden). He has authrored innumerable books and articles on different aspects of the English language, presented over 30 research papers at national and international conferences, and travelled widely. Prof. Gupta has guided several research dissertations in English. His current research interests include Development of Soft Skills, Eco-criticism and Queer Theory.

Sumana Biswas holds an M.A.and M.Phil degree in English Literature from Rani Durgavati Vishwavidalaya. At present she is pursuing Ph.D in the same University working on "Ecocritical Evaluation of Shakespeare's Comedies and Tragedies". Her publications include "Myriad Hues of Protest and Tumultuous Aftermath in V.S. Naipaul's Guerrillas" Her fields of interest are Shakespearean Literature, Contemporary Literature, Indian Literature, Feminism, Eco-feminism, Ecocriticism, Colonialism, Post colonialism, and other Socio-critical theories.

Wanda Baxter has a Master of Arts in English from the University of New Brunswick and a second Masters degree in Environmental Design (policy and planning) from the University of Calgary (2001). She has worked as an environmental planner, a language teacher, and as an organic gardener and farmer. She is preparing to do her interdisciplinary doctoral work to further explore the bridge between the fields of literature and the environment. Wanda has recently completed a PhD course in Teaching and Learning for Higher Education at Dalhousie University, and is interested in developing curriculum designed to teach Sustainability. She lives and works on a 17-acre farm just north of Lunenburg, Nova Scotia, Canada.

Ujjwal Jana is presently an Assistant Professor in the Department of English, Pondicherry University. He is the recipient of Fulbright Visiting Lecturer award (Indiana University, Bloomington, USA) 2007-2008. His interests include Indian Poetics, Indian Literature, Comparative Literature,

Translation Studies, and Marginal literature. He has published articles in some of the leading journals in the country and abroad. He has translated some poems for the Dalit Writings from Eastern India to be published by Oxford University Press. He is the co-editor of two books on Eco-critical Literature and one on Subaltern Vision in Postcolonial Indian English text to be published by Routledge, London.

Index